Test Yourself

Basic Mathematics

with Pre-Algebra

Patricia J. Newell, M.S.
Edison Community College
Fort Myers, FL

Contributing Editors

Shared Keny, Ph.D.
Department of Mathematics
Whittier College
Whittier, CA

Tony Julianelle, Ph.D.
Department of Mathematics
University of Vermont
Burlington, VT

Douglas G. Smith
Arthur P. Schalick High School
Elmer, NJ

NTC LearningWorks
NTC/Contemporary Publishing Group

Library of Congress Cataloging-in-Publication Data

Newell, Patricia J.
    Basic mathematics with pre-algebra / Patricia J. Newell ; contributing
editors, Shared Keny, Tony Julianelle, Douglas G. Smith.
      p.   cm. — (Test yourself)
   ISBN 0-8442-2351-4 (alk. paper)
    1. Mathematics—Examinations, questions, etc.   I. Keny, Shared.
II. Julianelle, Tony.   III. Smith, Douglas G. (Douglas George)
IV. Title.   V. Series: Test yourself (Lincolnwood, Ill.)
QA43.N42   1996
513'.12'076—dc21   96-47152
                                                  CIP

A *Test Yourself Books, Inc.* Project

Published by NTC Learning Works
An imprint of NTC/Contemporary Publishing Company
4255 West Touhy Avenue, Lincolnwood (Chicago), Illinois 60646-1975 U.S.A.
Copyright © 1996 by NTC/Contemporary Publishing Company
All rights reserved. No part of this book may be reproduced, stored in a retrieval
system, or transmitted in any form or by any means, electronic, mechanical,
photocopying, recording, or otherwise, without the prior permission of
NTC/Contemporary Publishing Company.
Printed in the United States of America
International Standard Book Number: 0-8442-2351-4
  6 7 8 9 0 VLP VLP 0 5 4 3

Contents

Preface

These test questions and answers were written for students and other adults who need to test their understanding of the concepts of basic mathematics and pre-algebra. Following each answer you will find a topic referral telling you where to locate more detailed explanations and problems in your textbook.

It is important for you to review the topics of the items you miss. Be sure to check your textbook or consult your instructor for further explanation. Just remember, your brain will only remember what it did. If you did a problem incorrectly and you didn't go back and correct it, you will probably do it incorrectly on your test.

Writing hundreds of questions and answers is a time-consuming project. I would like to thank Dana DeBellis for helping me get the first draft ready to submit. Thanks to Joan VanGlabek who is a great typist but an even better friend. And finally, a big thank you to Katie and Carol who made it possible for me to work on the weekends by doing all of the work around the house.

Patricia Newell, M.S.

How to Use this Book

This "Test Yourself" book is part of a unique series designed to help you improve your test scores on almost any type of examination you will face. Too often, you will study for a test—quiz, midterm, or final—and come away with a score that is lower than anticipated. Why? Because there is no way for you to really know how much you understand a topic until you've taken a test. The *purpose* of the test, after all, is to test your complete understanding of the material.

The "Test Yourself" series offers you a way to improve your scores and to actually test your knowledge at the time you use this book. Consider each chapter a diagnostic pretest in a specific topic. Answer the questions, check your answers, and then give yourself a grade. Then, and only then, will you know where your strengths and, more important, weaknesses are. Once these areas are identified, you can strategically focus your study on those topics that need additional work.

Each book in this series presents a specific subject in an organized manner, and although each "Test Yourself" chapter may not correspond to exactly the same chapter in your textbook, you should have little difficulty in locating the specific topic you are studying. Written by educators in the field, each book is designed to correspond, as much as possible, to the leading textbooks. This means that you can feel confident in using this book, and that regardless of your textbook, professor, or school, you will be much better prepared for anything you will encounter on your test.

Each chapter has four parts:

Brief Yourself. All chapters contain a brief overview of the topic that is intended to give you a more thorough understanding of the material with which you need to be familiar. Sometimes this information is presented at the beginning of the chapter, and sometimes it flows throughout the chapter, to review your understanding of various *units* within the chapter.

Test Yourself. Each chapter covers a specific topic corresponding to one that you will find in your textbook. Answer the questions, either on a separate page or directly in the book, if there is room.

Check Yourself. Check your answers. Every question is fully answered and explained. These answers will be the key to your increased understanding. If you answered the question incorrectly, read the explanations to *learn* and *understand* the material. You will note that at the end of every answer you will be referred to a specific subtopic within that chapter, so you can focus your studying and prepare more efficiently.

Grade Yourself. At the end of each chapter is a self-diagnostic key. By indicating on this form the numbers of those questions you answered incorrectly, you will have a clear picture of your weak areas.

There are no secrets to test success. Only good preparation can guarantee higher grades. By utilizing this "Test Yourself" book, you will have a better chance of improving your scores and understanding the subject more fully.

Whole Numbers

Brief Yourself

This chapter covers the basic mathematical skills used to simplify complicated problems. It contains questions and answers about place value, basic arithmetic operations (addition, subtraction, multiplication, division), properties of addition and multiplication, word problems, exponents, order of operations, and equations containing whole numbers.

Numbers are represented by symbols called numerals. Numerals may consist of one or more of the digits 0, 1, 2, 3, 4, 5, 6, 7, 8, and 9, which are used in the decimal number system. A whole number is a member of the set $\{0, 1, 2, 3, 4 \ldots\}$. The position or place value of a digit indicates the value that the digit represents. The place values are grouped into sets of three and separated by commas. Each position has a value ten times that of the place value to its right.

hundred millions	ten millions	millions,	hundred thousands	ten thousands	thousands,	hundreds	tens	ones
2	7,	4	0	5,	2	1	6

IN WORDS   "Twenty-seven million, four hundred five thousand, two hundred sixteen"

EXPANDED FORM $20{,}000{,}000 + 7{,}000{,}000 + 400{,}000 + 5{,}000 + 200 + 10 + 6$

Commutative Property of Addition — Changing the order of the addends does not change the sum.
$a + b = b + a$

Associative Property of Addition — Changing the grouping of the addends does not change the sum.
$(a + b) + c = a + (b + c)$

Additive Identity — Zero is the additive identity. The sum of any number and zero is that number. $a + 0 = a$

Commutative Property of Multiplication — Changing the order of the factors does not change the product.
$(a)(b) = (b)(a)$

Associative Property of Multiplication — Changing the grouping of the factors does not change the product.
$(a \cdot b)c = a(b \cdot c)$

Multiplicative Identity — The number 1 is called the multiplicative identity. The product of any number and 1 is that number. $(a)(1) = a$

Multiplication Property of 0 — The product of any number and 0 is always 0. $(a)\,(0) = 0$

Distributive Property — $a\,(b + c) = a \cdot b + a \cdot c$ or $(b + c)\,a = b \cdot a + c \cdot a$

Key Words Used in Word Problems

ADDITION	SUBTRACTION	MULTIPLICATION	DIVISION
plus	minus	times	divide
added to	subtracted from	multiplied by	divided by
increased by	difference	product of	divided into
more than	deduct	of	quotient
total	remains		each
sum			per
and			

Exponents are used to write repeated multiplications of the same factor. The exponent indicates the number of times the base is used as a factor.

$5^4 = (5)\,(5)\,(5)\,(5) = 625$ "Five to the fourth power equals 625"

Exponent Rules

$x^a \cdot x^b = x^{a+b}$ Multiplying like bases — leave the base the same and add the exponents.

$(x^a)^b = x^{(a)\,(b)}$ Raising a power to a power — leave the base the same and multiply the exponents.

$(x \cdot y)^a = x^a \cdot y^a$ The power of a product is the product of the powers.

$\dfrac{x^a}{x^b} = x^{a-b}$ Dividing like bases — leave the base the same and subtract the exponents.

$\left(\dfrac{x}{y}\right)^a = \dfrac{x^a}{y^a}$ Raising a fraction to a power — raise the numerator and denominator to the power.

$x^0 = 1$ Any non-zero number raised to the zero power is equal to 1.

$x^1 = x$ Any number raised to the first power is equal to that number.

Order of Operations — When evaluating a mathematical expression, perform the operations in the following order:

1) If the expression contains a grouping symbol such as ( ), [ ], { }, or a fraction bar, perform the operations inside the grouping symbols or above or below the fraction bar.
2) Evaluate any numbers with exponents.
3) Do all multiplication and divisions in the order that they appear, going from left to right.
4) Do all additions and subtractions in the order they appear, going from left to right.

Solving Equations with Whole Numbers — To solve an equation means to find the value of the unknown that makes the equation true. This is done by isolating the unknown on one side of the equation.

$x + a = b$ Subtract a from both sides.

$x - a = b$ Add a to both sides.

$ax = b$ Divide both sides by a.

$\dfrac{x}{a} = b$ Multiply both sides by a.

$ax + b = c$ Subtract the b first, then divide by a, using the rules above.

Test Yourself

1. Name the position of the 5 in each numeral.

   a. 8$\underline{5}$2

   b. 6$\underline{5}$,231

   c. 8,21$\underline{5}$

   d. 2$\underline{5}$6,193

2. Tell the value of each underlined digit.

   a. 3,$\underline{7}$26

   b. $\underline{4}$,263,916

   c. 73,$\underline{5}$29

   d. 5,$\underline{9}$27,364

3. Write each of the following as a numeral.

   a. fifteen thousand, three hundred twelve

   b. one million, four hundred sixty-two thousand, three hundred twenty-four

   c. two hundred thousand, five hundred

   d. three hundred eight thousand

4. Give the word name for each of the following.

   a. 23,567

   b. 200,304

   c. 2,405

   d. 27,602,158

5. Write each of the following in expanded form.

   a. 352

   b. 20,721

   c. 352,000

   d. 5,126

In problems 6 – 10, add.

6. 638 + 275

7. 534 + 4 + 6 + 2,322

8. 21,134 + 43,042

9. 2,756 + 39 + 872

10. 282 + 408 + 1,995 + 3 + 187 + 3,586

In problems 11 – 15, subtract.

11. 106 – 77

12. 978 – 586

13. 23,000 – 5,296

14. 59,736 – 3,219

15. 2,701 – 984

In problems 16 – 20, multiply.

16. 42 · 78

17. 6,587 · 9

18. 56 · 673

19. 639 · 358

20. 237 · 500

In problems 21 – 25, divide.

21. $8008 \div 26$

22. $61,825 \div 205$

23. $15,998 \div 38$

24. $3,871 \div 763$

25. $341,575 \div 815$

In problems 26 – 35, identify the property of addition or multiplication that is illustrated.

26. $(3)(5)(2) = (3)(2)(5)$

27. $4(3 + 2) = 4(3) + 4(2)$

28. $6 \cdot 1 = 6$

29. $7 + (11 + 2) = (7 + 11) + 2$

30. $0 + 9 = 9$

31. $(8 + 6) + 5 = (6 + 8) + 5$

32. $(4 \cdot 2)3 = 4(2 \cdot 3)$

33. $6(2 + 8) = (2 + 8)6$

34. $3 \cdot 0 = 0$

35. $(3 + 4) + 5 = 4 + (3 + 5)$

36. Alaska is the largest state in the United States, and Rhode Island is the smallest. The area of Alaska is 586,000 square miles, and Rhode Island has an area of 1,200 square miles. How much bigger is Alaska than Rhode Island in square miles? What is the total number of square miles in both states together?

37. The local car dealer is offering a $1,200 rebate on all cars that cost more than $18,000. What is the final cost of a car that costs $21,895?

38. A company distributes $21,312 in year-end bonuses. If each of the 36 employees receives the same amount, what bonus will each receive?

39. Katie borrows $1,300 and is charged $332 in interest. If she pays off the loan and the interest in monthly payments over a one-year period, how much will she have to pay per month?

40. Carol worked five months with a salary of $1,260 per month. She then received a $96 per month raise and continued to work at that rate for the rest of the year. What was her total yearly income?

41. Norma owns four rental units. They rent for $525 per month, and all four are occupied for an entire year. What is Norma's income from the four rental units?

42. There are 3,582 tickets available for a charity ball, and all but 379 have been sold. How many tickets have been sold?

43. A rectangle has four sides, two equal lengths and two equal widths. Find the perimeter (distance around) of a rectangle whose length is 24 inches and whose width is 47 inches.

44. Sarah buys a used car on the following terms: She pays $2,500 as a down payment and then $124 per month for 36 months. What is the total cost of her car?

45. A car travels 1,008 miles and uses 42 gallons of gasoline. How many miles per gallon does the car average?

Evaluate each of the following.

46. 2^4

47. 3^5

48. 1^7

49. 8^3

50. 6^0

Simplify each of the following.

51. $3^{12} \cdot 3^8$

52. $(9x^4)^2$

53. $(3^4)^5$

54. $(x^2)^4$

55. $(2 \cdot 5)^3$

56. $(2x^3y^4)^5$

57. $\dfrac{4^{12}}{4^9}$

58. $\dfrac{y^8}{y^3}$

59. $\dfrac{3^4 \cdot 3^7}{3^3}$

60. $\dfrac{5^{10}}{5^2 \cdot 5}$

61. $(4x^3y)^2 (2x^2y^5)^3$

62. $(4x^3) (6x^5) (2x^{10})$

63. $(x \cdot y)^8$

64. $(7x^3y^2) (8xy^3)$

65. $\dfrac{x^7 \cdot x^4}{x^2 \cdot x^3}$

Evaluate each of the following.

66. $(4)(7) - (2)(5)$

67. $5 + 2(7 - 1)$

68. $9 \cdot 2^3 + 36 \div 3^2 - 8$

69. $3 + 2[20 - 3(5 - 2)]$

70. $15 - 3(9 - 7)$

71. $6 + 3[9 - 2(4 - 1)]$

72. $5 \cdot 3^4 + 16 \div 8 - 2^2$

73. $3 + 2 \cdot 5^2$

74. $7(6)^2 - 6(2)^3$

75. $3(2 + 6 \cdot 8)$

76. $6 + [3^2 - (5 - 2)]$

Solve each of the following equations.

77. $x + 4 = 10$

78. $x - 3 = 7$

79. $8x = 72$

80. $\dfrac{x}{5} = 10$

81. $5x + 7 = 12$

82. $3x - 4 = 5$

83. $x + 3 = 37$

84. $x - 9 = 11$

85. $\dfrac{x}{8} - 5 = 2$

Check Yourself

1. a. The five is in the tens place.

   b. The five is in the thousands place.

   c. The five is in the ones place. This is sometimes referred to as the units place.

   d. The five is in the ten-thousands place. **(Place values)**

2. a. 700   Since the 7 is in the hundreds place, it has a value of (7)(100).

   b. 4,000,000   Since the 4 is in the millions place, it has a value of (4)(1,000,000).

   c. 3,000   Since the 3 is in the thousands place, it has a value of (3)(1000).

   d. 900,000   Since the 9 is in the hundred thousands place, it has a value of (9)(100,000). **(Place values)**

3. a. 15,312

   b. 1,462,324

   c. 200,500

   d. 308,000 **(Place values)**

4. a. Twenty-three thousand, five hundred sixty-seven

   b. Two hundred thousand, three hundred four

   c. Two thousand, four hundred five

   d. Twenty-seven million, six hundred two thousand, one hundred fifty-eight **(Place values)**

5 a. 300 + 50 + 2

   b. 20,000 + 700 + 20 + 1

   c. 300,000 + 50,000 + 2,000

   d. 5,000 + 100 + 20 + 6 **(Place values)**

Problems 6 – 10 are addition problems. When adding it is important to arrange the addends so that the digits that are in the same place value are in the same columns. Once the numbers are arranged properly, add using the basic addition facts.

6.
$$
\begin{array}{r}
638 \\
\underline{275} \\
913
\end{array}
$$

   (Whole number operations)

7.
$$
\begin{array}{r}
534 \\
4 \\
6 \\
\underline{2,322} \\
2,866
\end{array}
$$

   (Whole number operations)

8.
$$
\begin{array}{r}
21,134 \\
\underline{43,042} \\
64,176
\end{array}
$$

   (Whole number operations)

9.
$$
\begin{array}{r}
2,756 \\
39 \\
\underline{872} \\
3,667
\end{array}
$$

   (Whole number operations)

10.
```
     289
     408
   1,995
       3
     187
   3,586
   ─────
   6,461
```

(Whole number operations)

Problems 11 – 15 are subtraction problems. When a problem is initially written horizontally, it is easier to do the problem if you rewrite it vertically first. When a larger digit is being subtracted from a smaller digit, it is necessary to borrow. Remember that every place value is worth ten times the place value to its right. For example, when you borrow 1 from the hundreds, it becomes 10 tens.

11.
```
   106
    77
   ───
    29
```
(Whole number operations)

12.
```
   978
   586
   ───
   392
```
(Whole number operations)

13.
```
   23,000
    5,296
   ──────
   17,704
```
(Whole number operations)

14.
```
   59,736
    3,219
   ──────
   56,517
```
(Whole number operations)

15.
```
   2,701
     984
   ─────
   1,717
```
(Whole number operations)

Problems 16 – 20 are multiplication problems. It is easier to do these problems if they are written with the longer number on the top. See problem 18 as an example. As the multiplication is done with each digit of the multiplier, the partial product should be started in the same column as that digit.

16.
```
     42
     78
   ────
    336     Multiply 42 by 8.
   2940      Multiply 42 by 7. Remember that the 7 really means 70 since it is in the tens place.
   ─────     However, it is not necessary to place the 0 on this line.
   3,276
```

(Whole number operations)

17.
```
   6,587
       9
   ──────
   59,283
```

(Whole number operations)

18.
```
      673
       56
     4038
    33650
    37,688
```

(Whole number operations)

19.
```
       639
       358
     5 112
    31 950
   191 700
   228,762
```

(Whole number operations)

20.
```
       237
       500
       000
     0 000
   118 500
   118,500
```

(Whole number operations)

Problems 21 – 25 are division problems. The order of steps to be followed is: divide, multiply, subtract, bring down the next digit, then start the steps over. If it is not possible to divide after a number is brought down, remember to put a 0 in the quotient.

21.
```
         308
    26)8008          Divide 26 into 80.
       78            Multiply 3 times 26.
       20            Subtract, then bring down the next digit, which is a 0.
        0            Divide 26 into 20.
      208            Multipy 0 times 26, subtract, bring down the next digit which is an 8.
      208            Divide 26 into 208.
        0
```

(Whole number operations)

22.
```
          3 01 R120
    205 )61825
        615
         32
          0
        325
        205
        120
```

(Whole number operations)

23.
$$\begin{array}{r} 421 \\ 38\overline{)15998} \\ \underline{152} \\ 79 \\ \underline{76} \\ 38 \\ \underline{38} \\ 0 \end{array}$$

(Whole number operations)

24.
$$\begin{array}{r} 5 \text{ R}56 \\ 763\overline{)3871} \\ \underline{3815} \\ 56 \end{array}$$

(Whole number operations)

25.
$$\begin{array}{r} 419 \text{ R}90 \\ 815\overline{)341575} \\ \underline{3260} \\ 1557 \\ \underline{815} \\ 7425 \\ \underline{7335} \\ 90 \end{array}$$

(Whole number operations)

26. Commutative Property of Multiplication. The order of the factors has been changed.
     (Properties of arithmetic)

27. Distributive Property. The 4 multiplies both addends. **(Properties of arithmetic)**

28. Multiplicative Identity. Multiplying by 1 does not change the value of the number.
     (Properties of arithmetic)

29. Associative Property of Addition. The grouping of the addends has been changed.
     (Properties of arithmetic)

30. Additive Identity. Adding 0 does not change the value of the number. **(Properties of arithmetic)**

31. Commutative Property of Addition. The order of the addends has been changed. **(Properties of arithmetic)**

32. Associative Property of Multiplication. The grouping of the factors has been changed.
     (Properties of arithmetic)

33. Commutative Property of Multiplication. The position of the 6 has been changed.
     (Properties of arithmetic)

34. Multiplication Property of 0. Any number multiplied by 0 is equal to 0. **(Properties of arithmetic)**

35. Commutative and Associative Properties of Addition. Both the order and the grouping of the addends have
     been changed. **(Properties of arithmetic)**

36. 584,800 sq. miles — To find the difference of the sizes of the two states, subtraction should be used.

   587,200 sq. miles — To find the total area of the two states, addition should be used. **(Word problems)**

37. $20,695 — A rebate is an amount that is subtracted from the original price of the car. The only purpose for the $18,000 is to make sure the car qualifies for a rebate. It is not used in any calculation. **(Word problems)**

38. $592 — The money is to be divided into 36 equal parts. This requires the division of $21,312 by 36. **(Word problems)**

39. $136 — The total to be repaid is equal to the amount borrowed and the interest added together. Since there are 12 months in a year, it is then necessary to divide this total by 12 to find the monthly payment. **(Word problems)**

40. $15,792 — Multiply 5 times $1,260 to find the amount of money earned for the first 5 months. Add $96 to $1,260 to find the new monthly income. Multiply this by 7, the number of months remaining in the year. Add these two amounts together to find the total income for the year. **(Word problems)**

41. $25,200 — Multiply $525 times 12 (the number of months in a year). This will give the total amount of rent collected for each unit. This number needs to be multiplied by 4 to determine the total income collected from all of the units together. **(Word problems)**

42. 3203 tickets — Subtract 379 (the number of remaining tickets) from the number of tickets available. This will equal the number of tickets sold. **(Word problems)**

43. 142 inches — To find the perimeter, add the lengths of all of the sides. Since this is a rectangle, two lengths and two widths are needed. **(Word problems)**

44. $6,964 — To find the amount paid over the 36-month period, multiply $124 times 36. This amount needs to be added to the down payment. **(Word problems)**

45. 24 miles/gallon — To find the number of miles per gallon, divide the number of gallons used into the number of miles driven. **(Word problems)**

46. 16 — Two used as a factor 4 times (2)(2)(2)(2) **(Exponents)**

47. 243 — Three used as a factor 5 times (3)(3)(3)(3)(3) **(Exponents)**

48. 1 — One used as a factor 7 times (1)(1)(1)(1)(1)(1)(1) **(Exponents)**

49. 512 — Eight used as a factor 3 times (8)(8)(8) **(Exponents)**

50. 1 — Any non-zero number raised to the 0 power is equal to 1. **(Exponents)**

51. 3^{20} — When multiplying like bases, leave the base the same and add the exponents. **(Exponents)**

52. $81x^8$ — All factors in the parentheses need to be raised to the second power. When raising a power to a power, leave the base the same and multiply the exponents. **(Exponents)**

53.	3^{20}	To simplify a power raised to a power, leave the base the same and multiply the exponents. **(Exponents)**
54.	x^8	When raising a power to a power, leave the base the same and multiply the exponents. **(Exponents)**
55.	$2^3 \cdot 5^3 = 1000$	Both factors get raised to the third power. **(Exponents)**
56.	$32x^{15}y^{20}$	All factors in the parentheses need to be raised to the fifth power. When raising a power to a power, leave the base the same and multiply the exponents. **(Exponents)**
57.	4^3	When dividing like bases, the base stays the same and the exponents are subtracted (top minus bottom). **(Exponents)**
58.	y^5	When dividing like bases, leave the base the same and subtract the exponents (top minus bottom). **(Exponents)**
59.	3^8	Multiply the factors in the numerator by leaving the base the same and adding the exponents. Then divide the numerator and the denominator by leaving the base the same and subtracting the exponents. **(Exponents)**
60.	5^7	Multiply the factors in the denominator by leaving the base the same and adding the exponents. Then divide the numerator and the denominator by leaving the base the same and subtracting the exponents. **(Exponents)**
61.	$128x^{12}y^{17}$	Raise the factors inside the first parentheses to the second power and the factors in the second parentheses to the third power. Then multiply the like bases by adding the exponents. **(Exponents)**
62.	$48x^{18}$	When multiplying like bases, leave the base the same and add the exponents. Note that the 4, 6, and 2 are not exponents and therefore should be multiplied. **(Exponents)**
63.	x^8y^8	All factors in the parentheses need to be raised to the eighth power. **(Exponents)**
64.	$56x^4y^5$	When multiplying like bases, leave the base the same and add the exponents. Note that the 7 and the 8 are not exponents and therefore should be multiplied. **(Exponents)**
65.	x^6	Multiply the factors in the numerator by leaving the base the same and adding the exponents. Multiply the factors in the denominator by leaving the base the same and adding the exponents. Divide the numerator and the denominator by leaving the base the same and subtracting the exponents. **(Exponents)**
66.	$(4)(7) - (2)(5)$	
	$= 28 - 10$	Multiply as you move from left to right.
	$= 18$	Subtract. **(Order of operations)**

67. $5 + 2(7 - 1)$

    $= 5 + 2(6)$ Evaluate what is in the grouping symbols.

    $= 5 + 12$ Multiply before adding.

    $= 17$ Add. **(Order of operations)**

68. $9 \cdot 2^3 + 36 \div 3^2 - 8$

    $= 9 \cdot 8 + 36 \div 9 - 8$ Evaluate the exponents first.

    $= 72 + 4 - 8$ Multiply and divide in the order they appear going from left to right.

    $= 76 - 8$ Add.

    $= 68$ Subtract. **(Order of operations)**

69. $3 + 2[20 - 3(5 - 2)]$

    $= 3 + 2[20 - 3(3)]$ Evaluate the inside set of parentheses.

    $= 3 + 2[20 - 9]$ Multiply inside the brackets.

    $= 3 + 2[11]$ Subtract inside the brackets.

    $= 3 + 22$ Multiply.

    $= 25$ Add. **(Order of operations)**

70. $15 - 3(9 - 7)$

    $= 15 - 3(2)$ Subtract inside the parentheses.

    $= 15 - 6$ Multiply.

    $= 9$ Subtract. **(Order of operations)**

71. $6 + 3[9 - 2(4 - 1)]$

    $= 6 + 3[9 - 2(3)]$ Evaluate the inside set of parentheses.

    $= 6 + 3[9 - 6]$ Multiply inside the brackets.

    $= 6 + 3[3]$ Subtract inside the brackets.

    $= 6 + 9$ Multiply.

    $= 15$ Add. **(Order of operations)**

72. $5 \cdot 3^4 + 16 \div 8 - 2^2$

    $= 5 \cdot 81 + 16 \div 8 - 4$ Evaluate the exponents.

    $= 405 + 2 - 4$ Multiply and divide in the order they appear, going from left to right.

    $= 407 - 4$ Add.

    $= 403$ Subtract. **(Order of operations)**

73. $3 + 2 \cdot 5^2$

    $= 3 + 2 \cdot 25$ Evaluate the exponent.

	$= 3 + 50$	Multiply.
	$= 53$	Add. **(Order of operations)**
74.	$7\,(6)^{\,2} - 6\,(2)^{\,3}$	
	$= 7(36) - 6(8)$	Evaluate the exponents.
	$= 252 - 48$	Multiply. Since these multiplications do not affect each other, they can be done at the same time.
	$= 204$	Subtract. **(Order of operations)**
75.	$3\,(2 + 6 \cdot 8)$	
	$= 3(2 + 48)$	Multiply inside the parentheses.
	$= 3(50)$	Add inside the parentheses.
	$= 150$	Multiply. **(Order of operations)**
76.	$6 + [\,3^2 - (5 - 2)\,]$	
	$= 6 + [\,3^2 - (3)]$	Simplify inside the parentheses.
	$= 6 + [9 - 3]$	Evaluate the exponent inside the brackets.
	$= 6 + [6]$	Subtract inside the brackets.
	$= 12$	Add. **(Order of operations)**
77.	$x + 4 \;=\; 10$	
	$x + 4 - 4 \;=\; 10 - 4$	Subtract 4 from both sides.
	$x \;=\; 6$	
	(Equations)	
78.	$x - 3 \;=\; 7$	
	$x - 3 + 3 \;=\; 7 + 3$	Add 3 to both sides.
	$x \;=\; 10$	
	(Equations)	
79.	$\dfrac{8x}{8} = \dfrac{72}{8}$	Divide both sides by 8.
	$x \;=\; 9$	
	(Equations)	

80. $5 \cdot \dfrac{x}{5} = 10 \cdot 5$ Multiply both sides by 5.

$x = 50$

(Equations)

81. $5x + 7 = 12$

$5x + 7 - 7 = 12 - 7$ Subtract 7 from both sides.

$\dfrac{5x}{5} = \dfrac{5}{5}$ Divide both sides by 5.

$x = 1$

(Equations)

82. $3x - 4 = 5$

$3x - 4 + 4 = 5 + 4$ Add 4 to both sides.

$\dfrac{3x}{3} = \dfrac{9}{3}$ Divide both sides by 3.

$x = 3$

(Equations)

83. $x + 3 = 37$

$x + 3 - 3 = 37 - 3$ Subtract 3 from both sides.

$x = 34$

(Equations)

84. $x - 9 = 11$

$x - 9 + 9 = 11 + 9$ Add 9 to both sides.

$x = 20$

(Equations)

85. $\dfrac{x}{8} - 5 = 2$

$\dfrac{x}{8} - 5 + 5 = 2 + 5$ Add 5 to both sides.

$8 \cdot \dfrac{x}{8} = 7 \cdot 8$ Multiply both sides by 8.

$x = 56$

(Equations)

Grade Yourself

Circle the question numbers that you had incorrect. Then indicate the number of questions you missed. If you answered more than three questions incorrectly, you need to focus on that topic. (If a topic has less than three questions and you had at least one wrong, we suggest you study that topic also. Read your textbook or a review book, or ask your teacher for help.)

Subject: Whole Numbers

Topic	Question Numbers	Number Incorrect
Place values	1, 2, 3, 4, 5	
Whole number operations	6, 7, 8, 9, 10, 11, 12, 13, 14, 15, 16, 17, 18, 19, 20, 21, 22, 23, 24, 25	
Properties of arithmetic	26, 27, 28, 29, 30, 31, 32, 33, 34, 35	
Word problems	36, 37, 38, 39, 40, 41, 42, 43, 44, 45	
Exponents	46, 47, 48, 49, 50, 51, 52, 53, 54, 55, 56, 57, 58, 59, 60, 61, 62, 63, 64, 65	
Order of operations	66, 67, 68, 69, 70, 71, 72, 73, 74, 75, 76	
Equations	77, 78, 79, 80, 81, 82, 83, 84, 85	

Fractions

Brief Yourself

This chapter contains a review of the rules for working with fractions. It includes questions and answers about prime factorization, LCD, operations with fractions, complex fractions, word problems, exponents, order of operations, and equations.

A prime number is any number greater than 1 that has exactly two factors—itself and 1.
   Examples: 2, 3, 5, 7, 11, 13, 17, 19, 23 . . .

A composite number is any number greater than 1 that is not prime.
   NOTE: 0 and 1 are neither prime nor composite.

Divisibility Rules — A whole number is evenly divisible by

   2 — if it ends in an even digit — 0, 2, 4, 6, 8

   3 — if the sum of its digits is divisible by 3

   5 — if it ends in a 0 or a 5

   6 — if it is divisible by both 2 and 3

   9 — if the sum of its digits is divisible by 9

   10 — if it ends in a 0

Prime Factorization means to write a composite number as a product of prime factors.
   Example $60 = 2 \cdot 30$

$$2 \cdot 2 \cdot 15$$

$$2 \cdot 2 \cdot 3 \cdot 5 \text{ — all numbers are prime}$$

$$2^2 \cdot 3 \cdot 5$$

Equivalent Fractions are fractions that may look different but actually have the same value. There are two ways to find equivalent fractions:

1. Building a fraction — multiply the numerator and the denominator of the fraction by the same non-zero number. The resulting fraction will be equivalent to the original fraction. This is because the original fraction was multiplied by a form of 1 ($a/a = 1$), which is the multiplicative identity.

2. Reducing a fraction — divide the numerator and the denominator of a fraction by any common factors. To find the common factors, write the numerator and the denominator in prime factorization form. Then divide out (cancel) the common factors. Remember $a/a = 1$. A fraction is in lowest terms when the numerator (top) and the denominator (bottom) have no factors in common other than the number 1.

Types of Fractions

Proper — numerator is smaller than the denominator. Such a fraction has a value less than 1.

Improper — numerator is larger than or equal to the denominator. It has a value greater than or equal to 1.

Mixed Number — the sum of a whole number and a fraction $(8 + \frac{2}{3})$ with the plus sign omitted. $8\frac{2}{3}$

To change a mixed number to an improper fraction, multiply the denominator by the whole number and add the numerator. This number becomes the numerator of the fraction. The denominator is the same as that of the original fraction.

To change an improper fraction to a mixed number, divide the denominator into the numerator. The number of times it goes in evenly becomes the whole number part of the mixed number, and any remainder becomes the numerator of the fraction part. The denominator is the same as that of the original fraction.

Reciprocal — The reciprocal of a proper or improper fraction is formed by interchanging the numerator and the denominator of the original fraction. The product of a number and its reciprocal is always 1. The only number that does not have a reciprocal is 0.

Operations with Fractions

Multiplication — To multiply two or more fractions:

— Change all mixed numbers to improper fractions.

— Factor the numerators and the denominators and divide out all of the common factors.

— Write the product of the numerators over the product of the denominators.

$$\frac{a}{b} \cdot \frac{c}{d} = \frac{a \cdot c}{b \cdot d}$$

Division — To divide two or more fractions:

— Change all mixed numbers to improper fractions.

— Change the division to multiplication and replace the second fraction with its reciprocal.

— Factor all the numerators and denominators and divide out all the common factors.

— Write the product of the numerators over the product of the denominators.

$$\frac{a}{b} \div \frac{c}{d} = \frac{a}{b} \cdot \frac{d}{c} = \frac{a \cdot d}{b \cdot c}$$

Like Fractions are fractions with common (or the same) denominators. Fractions with unlike (or different) denominators can be rewritten as equivalent fractions with common denominators.

LCD — Least Common Denominator — the smallest number that all of the given denominators will divide into evenly. The LCD cannot be smaller than the largest of the given denominators.

To find the LCD, write the given denominators in prime factorization form using exponents when necessary. The LCD will be the product of the highest power of each factor that appears in these prime factorizations.

Addition

To add fractions with like denominators:

— Add any whole numbers together.

— Add the fractions together by adding the numerators and keeping the common denominator.

— Reduce the answer to lowest terms. If the fraction is improper, write it as a mixed number and then combine the whole number parts.

To add fractions with unlike denominators:

— Find the LCD for the given fraction.

— Rewrite the fractions as equivalent fractions using the LCD as the new denominator.

— To finish the problem, use the rules for like denominators.

— Reduce the answer if possible.

$$\frac{a}{b} + \frac{c}{b} = \frac{a+c}{b} \qquad \text{or} \qquad \frac{a}{b} + \frac{c}{d} = \frac{(a)\,(d)}{(b)\,(d)} + \frac{(c)\,(b)}{(d)\,(b)} = \frac{(a)\,(d) + (c)\,(d)}{(b)\,(d)}$$

Subtraction — follows the same rules, except you subtract the whole numbers and subtract the numerators.

$$\frac{a}{b} - \frac{c}{b} = \frac{a-c}{b} \qquad \text{or} \qquad \frac{a}{b} - \frac{c}{d} = \frac{(a)\,(d)}{(b)\,(d)} - \frac{(c)\,(b)}{(d)\,(b)} = \frac{(a)\,(d) - (c)\,(b)}{(b)\,(d)}$$

The only situation in subtraction that requires any further discussion occurs when borrowing is involved. If this is necessary, rename the minuend (the top or first mixed number) so that it is one smaller in the whole number part and the equivalent of one more in the fraction part.

Example: $7\frac{2}{5} = 6 + \frac{5}{5} + \frac{2}{5} = 6 + \frac{7}{5} = 6\frac{7}{5}$

Once this has been completed, proceed as with other subtraction problems. Remember, an LCD is still needed and should be found before you borrow or rename.

NOTE: If all of the mixed numbers are changed to improper fractions, borrowing is unnecessary. An LCD is still required.

NOTE: All problems that include adding and subtracting of mixed numbers can also be done by first changing all of the mixed numbers to improper fractions.

To compare the size of two or more fractions, rewrite them as fractions with a common denominator. The larger the numerator, the larger the value of the fraction.

A complex fraction is a fraction whose numerator and/or denominator are fractions themselves.

To simplify a complex fraction:

— First perform any operations that are in the numerator and/or the denominator.

— Rewrite the problem as a division problem.

— Follow the rules of division as stated above.

Order of Operations — Combine the rules for order of operations with the rules of fractions.

Equations — Combine the rules of equations with the rules of fractions.

Recall that $bx = c$

$$\frac{bx}{b} = \frac{c}{b}$$

$$x = \frac{c}{b}$$

When b is a fraction, it is easier to multiply both sides by the reciprocal of b than it is to divide by b. Dividing is the same as multiplying by the reciprocal.

When solving word problems, refer to the key word list in Chapter 1 to determine the appropriate operation to use and then incorporate the rules of fractions.

Test Yourself

1. Is 22,390 divisible by

   a. 2?

   b. 3?

   c. 5?

   d. 6?

   e. 9?

   f. 10?

2. Write the prime factorization of each of the following.

   a. 108

   b. 84

   c. 90

   d. 51

3. Replace the ? to make a true statement.

   a. $\frac{3}{5} = \frac{?}{20}$

   b. $\frac{5}{8} = \frac{?}{72}$

   c. $\frac{11}{12} = \frac{5951}{?}$

   d. $\frac{11x}{13} = \frac{?}{52}$

   e. $\frac{?}{56} = \frac{3y}{14}$

   f. $\frac{7}{15p} = \frac{?}{75pq}$

4. Reduce each of the following to lowest terms.

   a. $\frac{45}{75}$

   b. $\frac{64}{72}$

   c. $\frac{90}{126}$

d. $\dfrac{6x}{10x}$

e. $\dfrac{24xy}{40y}$

f. $\dfrac{150ab^2}{210ab}$

5. Pick the largest fraction from each group.

   a. $\dfrac{7}{14}$ $\dfrac{5}{8}$

   b. $\dfrac{2}{5}$ $\dfrac{12}{35}$

   c. $\dfrac{7}{12}$ $\dfrac{5}{9}$

   d. $\dfrac{3}{5}$ $\dfrac{7}{15}$ $\dfrac{28}{45}$

   e. $\dfrac{5}{7}$ $\dfrac{13}{14}$ $\dfrac{10}{21}$

6. Insert <, =, or > to make each statement true.

   a. $\dfrac{3}{7}$ $\dfrac{4}{9}$

   b. $\dfrac{7}{12}$ $\dfrac{3}{5}$

   c. $\dfrac{5}{18}$ $\dfrac{1}{4}$

   d. $\dfrac{18}{27}$ $\dfrac{2}{3}$

7. Change each mixed number to an equivalent improper fraction.

   a. $4\dfrac{7}{16}$

   b. $8\dfrac{2}{3}$

   c. $10\dfrac{1}{2}$

   d. $17\dfrac{4}{5}$

8. Change each improper fraction to an equivalent mixed number.

   a. $\dfrac{49}{11}$

b. $\dfrac{17}{2}$

c. $\dfrac{193}{9}$

d. $\dfrac{140}{20}$

Multiply each of the following. Be sure all answers are reduced to lowest terms.

9. $\left(\dfrac{8}{10}\right)\left(\dfrac{3}{4}\right)\left(\dfrac{2}{9}\right)$

10. $\left(\dfrac{4}{3}\right)\left(\dfrac{9}{20}\right)$

11. $\left(\dfrac{3}{5}\right)(15)$

12. $\left(\dfrac{72}{35}\right)\left(\dfrac{55}{108}\right)\left(\dfrac{7}{110}\right)$

13. $\left(4\dfrac{1}{5}\right)\left(1\dfrac{1}{7}\right)$

14. $\left(5\dfrac{3}{8}\right)\left(3\dfrac{1}{2}\right)$

15. $\left(\dfrac{a^2b}{c}\right)\left(\dfrac{c^3}{ab^2}\right)$

16. $\left(\dfrac{12a^2b}{c^3}\right)\left(\dfrac{5c}{18ab}\right)$

Divide each of the following. Be sure all answers are reduced to lowest terms.

17. $\dfrac{5}{18} \div \dfrac{10}{27}$

18. $\dfrac{25}{46} \div \dfrac{40}{69}$

19. $16 \div \dfrac{1}{8}$

20. $\dfrac{3}{5} \div 2$

21. $\dfrac{7}{8} \div \dfrac{8}{7}$

22. $\dfrac{12a^2}{5b} \div \dfrac{6a}{5b^2}$

23. $\dfrac{x}{y^2} \div \dfrac{x^3}{y}$

24. $\dfrac{a^2 b}{c^2} \div \dfrac{a}{c^3}$

Add or subtract each of the following. Be sure all answers are reduced to lowest terms.

25. $\dfrac{5}{23} + \dfrac{7}{23} + \dfrac{9}{23}$

26. $\dfrac{12}{25} - \dfrac{7}{25}$

27. $12\dfrac{3}{4} + 5\dfrac{1}{4}$

28. $7\dfrac{11}{15} - 3\dfrac{8}{15}$

29. $13\dfrac{4}{15} + 12\dfrac{11}{15} + 26\dfrac{3}{15}$

30. $12 - 10\dfrac{11}{16}$

31. $8\dfrac{1}{6} - 2\dfrac{5}{6}$

32. $\dfrac{3}{8} + \dfrac{1}{12} + \dfrac{5}{24}$

33. $\dfrac{11}{12} - \dfrac{2}{5}$

34. $\dfrac{5}{24} + \dfrac{2}{24} - \dfrac{1}{24}$

35. $12\dfrac{7}{8} - 3\dfrac{5}{6}$

36. $\dfrac{7}{16} + \dfrac{3}{20} + \dfrac{1}{5}$

37. $12\dfrac{3}{10} - 5\dfrac{7}{10}$

38. $3 - \dfrac{5}{6}$

39. $12\dfrac{11}{12} + 18 + 8\dfrac{5}{8}$

40. $14\dfrac{2}{3} - 7$

41. $9\dfrac{2}{3} - 5\dfrac{3}{4}$

42. $\dfrac{19}{42} + \dfrac{13}{70}$

43. $13\dfrac{1}{6} - 12\dfrac{15}{24}$

Simplify each of the following. Be sure that the answer is reduced to lowest terms.

44. $\dfrac{\frac{1}{8}}{\frac{3}{4}}$

45. $\dfrac{\frac{2}{3}}{\frac{2}{7}}$

46. $\dfrac{\frac{14}{5}}{\frac{7}{5}}$

47. $\dfrac{\frac{3}{4} + \frac{2}{5}}{\frac{1}{2} + \frac{3}{5}}$

48. $\dfrac{\frac{7}{6} + \frac{2}{3}}{\frac{3}{2} - \frac{8}{9}}$

49. $\dfrac{\frac{3}{4}}{5 - \frac{1}{8}}$

50. $\dfrac{\dfrac{4}{5} + \dfrac{3}{10}}{\dfrac{7}{8} + \dfrac{3}{4}}$

51. $\dfrac{2 + \dfrac{3}{4}}{1 - \dfrac{1}{8}}$

Simplify each of the following. Be sure the answer is reduced to lowest terms.

52. $\left(\dfrac{7}{8} - \dfrac{1}{2}\right) \div \dfrac{3}{11}$

53. $2\dfrac{3}{8} + \dfrac{1}{2}\left(\dfrac{1}{3} + 1\dfrac{2}{3}\right)^3$

54. $2\left(\dfrac{1}{2} + \dfrac{1}{3}\right) + 3\left(\dfrac{2}{3} + \dfrac{1}{4}\right)$

55. $\left(\dfrac{3}{4} \div \dfrac{6}{5}\right) + \left(\dfrac{3}{4} \cdot \dfrac{6}{5}\right)$

56. $(32)\left(\dfrac{3}{4}\right)^2 - \dfrac{1}{2}(48)\left(\dfrac{3}{8}\right)$

57. $\left(\dfrac{3}{4}\right)(3)\left(\dfrac{5}{9}\right) \div \left(\dfrac{5}{8}\right)^2$

58. $3 - \dfrac{1}{8}\left[5\dfrac{1}{2} - 4\dfrac{7}{8}\right]^2$

59. $3\dfrac{1}{4} + 1\dfrac{1}{5}\left[\dfrac{1}{4} + 2\right]$

60. $\dfrac{9}{24} + \left[\dfrac{1}{8} + \left(\dfrac{1}{6}\right)^2\right]$

Solve each of the following. Be sure the answer is reduced to lowest terms.

61. $\dfrac{1}{3}x = 7$

62. $\dfrac{x}{6} + 1 = \dfrac{4}{3}$

63. $\dfrac{1}{3}a - \dfrac{4}{5} = \dfrac{8}{15}$

64. $\dfrac{3}{7} + \dfrac{1}{4}c = \dfrac{13}{28}$

65. $\dfrac{y}{5} + 2 = \dfrac{11}{5}$

66. $\dfrac{1}{6} + \dfrac{x}{4} = \dfrac{17}{12}$

67. $\dfrac{14}{15} = x + \dfrac{1}{30}$

68. $\dfrac{2}{9}y = 4$

69. $\dfrac{3}{7}x - \dfrac{4}{14} = \dfrac{1}{2}$

70. $\dfrac{8}{9}y + \dfrac{2}{3} = \dfrac{5}{6}$

Solve each of the following. Be sure the answer is reduced to lowest terms.

71. Bill needs a piece of ribbon $11\dfrac{3}{8}$ inches long for a pair of shorts he is making for his daughter Mindy. He also needs a piece $6\dfrac{2}{5}$ inches long for the matching shirt. What is the total length of ribbon that Bill needs?

72. Grandma has a recipe for "poonie cakes," an old family favorite. If she put 60 poonie cakes on the plate and Jimmy ate $\dfrac{2}{3}$ of them, how many poonie cakes did Jimmy eat?

73. Michelle promised her mother she would practice the violin 7 hours this week. If she has practiced $4\dfrac{3}{4}$ hours so far this week, how many more hours does Michelle need to practice?

74. Linda cut two pieces of fabric from a bolt of fabric containing 30 yards. The pieces she cut were $2\dfrac{3}{4}$ yards and $3\dfrac{2}{3}$ yards. How much fabric remains on the bolt?

75. How many pieces, each $\dfrac{3}{10}$ of an inch long, can be cut from piece of chain that is 6 inches long?

76. The distance around the track is $\frac{3}{8}$ of a mile. How many times must Carmen run around the track if she wants to run 27 miles?

77. A light pole is $86\frac{1}{4}$ inches long. If you want $61\frac{5}{8}$ inches of the pole above the ground, how far should the pole be set into the ground?

78. A roll of carpet had $23\frac{2}{3}$ yards of carpet on it. The salesman sold $\frac{3}{4}$ of the roll to a customer. How many yards of carpet did he sell?

79. A number is divided by 6. If the quotient is decreased by $\frac{5}{9}$, the result is $\frac{2}{3}$. What is the number?

80. Ethan read $\frac{1}{3}$ of a book on Monday and $\frac{1}{4}$ of the book on Tuesday. If the book has 312 pages, how many pages did Ethan read?

Check Yourself

1. a. Yes, it ends in an even digit, 0.

   b. No, the sum of its digits is 16, which is not divisible by 3.

   c. Yes, it ends in a 0.

   d. No, it must be divisible by both 2 and 3.

   e. No, the sum of its digits is 16, which is not divisible by 9.

   f. Yes, it ends in a 0.

   (Divisibility rules)

2. a. 108　You can start with any product that will give you 108. The final product will always be the same combination of prime factors, no matter how you start.

   $= 2 \cdot 54$

   $= 2 \cdot 2 \cdot 27$

   $= 2 \cdot 2 \cdot 3 \cdot 9$

   $= 2 \cdot 2 \cdot 3 \cdot 3 \cdot 3$

   $= 2^2 \cdot 3^3$

   b. 84

   $= 4 \cdot 21$

   $= 2 \cdot 2 \cdot 21$

$= 2 \cdot 2 \cdot 3 \cdot 7$

$= 2^2 \cdot 3 \cdot 7$

c. 90

$= 9 \cdot 10$

$= 3 \cdot 3 \cdot 10$

$= 3 \cdot 3 \cdot 2 \cdot 5$

$= 2 \cdot 3^2 \cdot 5$ It is standard to put the factors in ascending numerical order.

d. 51

$= 3 \cdot 17$ Use the divisibility rule to get started. The sum of the digits is 6, which is divisible by 3, which means 51 is divisible by 3.

(Prime factorization)

3. a. $? = 12$ Multiply both the numerator and the denominator by 4.

b. $? = 45$ Multiply both the numerator and the denominator by 9.

c. $? = 6{,}492$ Multiply both the numerator and the denominator by 541.

d. $? = 44x$ Multiply both the numerator and the denominator by 4.

e. $? = 12y$ Multiply both the numerator and the denominator by 4.

f. $? = 35q$ Multiply both the numerator and the denominator by $5q$.

(Equivalent fractions)

4. a. $\dfrac{3 \cdot 3 \cdot 5}{3 \cdot 5 \cdot 5} = \dfrac{3}{5}$ Prime factor and divide out (cancel) the common factors.

b. $\dfrac{2 \cdot 2 \cdot 2 \cdot 2 \cdot 2 \cdot 2}{2 \cdot 2 \cdot 2 \cdot 3 \cdot 3} = \dfrac{2 \cdot 2 \cdot 2}{3 \cdot 3} = \dfrac{8}{9}$

c. $\dfrac{2 \cdot 3 \cdot 3 \cdot 5}{2 \cdot 3 \cdot 3 \cdot 7} = \dfrac{5}{7}$

d. $\dfrac{2 \cdot 3 \cdot x}{2 \cdot 5 \cdot x} = \dfrac{3}{5}$

e. $\dfrac{2 \cdot 2 \cdot 2 \cdot 3 \cdot x \cdot y}{2 \cdot 2 \cdot 2 \cdot 5 \cdot y} = \dfrac{3 \cdot x}{5} = \dfrac{3x}{5}$

f. $\dfrac{2 \cdot 3 \cdot 5 \cdot 5 \cdot a \cdot b \cdot b}{2 \cdot 3 \cdot 5 \cdot 7 \cdot a \cdot b} = \dfrac{5 \cdot b}{7} = \dfrac{5b}{7}$

(Reducing fractions)

5. a. $\frac{5}{8}$ Change to equivalent fractions with a common denominator. The larger the numerator represents the

   larger fraction. $\frac{7}{14}$ and $\frac{5}{8}$ become $\frac{28}{56}$ and $\frac{35}{56}$.

   b. $\frac{2}{5}$ $\frac{2}{5}$ and $\frac{12}{35}$ become $\frac{14}{35}$ and $\frac{12}{35}$.

   c. $\frac{7}{12}$ $\frac{7}{12}$ and $\frac{5}{9}$ become $\frac{21}{36}$ and $\frac{20}{36}$.

   d. $\frac{28}{45}$ $\frac{3}{5}, \frac{7}{15}, \frac{28}{45}$ become $\frac{27}{45}, \frac{21}{45}, \frac{28}{45}$ respectively.

   e. $\frac{13}{14}$ $\frac{5}{7}, \frac{13}{14}, \frac{10}{21}$ become $\frac{30}{42}, \frac{39}{42}, \frac{20}{42}$ respectively.

   (Reducing fractions)

6. a. $<$ Change to equivalent fractions with an LCD and compare numerators.

   Thus, $\frac{27}{63}$ is compared to $\frac{28}{63}$.

   b. $<$ $\frac{35}{60}$ compared to $\frac{36}{60}$

   c. $>$ $\frac{10}{36}$ compared to $\frac{9}{36}$

   d. $=$ $\frac{18}{27}$ compared to $\frac{18}{27}$

   (Reducing fractions)

7. a. $\frac{71}{16}$ $16 \cdot 4 + 7 = 71$

   b. $\frac{26}{3}$ $3 \cdot 8 + 2 = 26$

   c. $\frac{21}{2}$ $2 \cdot 10 + 1 = 21$

   d. $\frac{89}{5}$ $5 \cdot 17 + 4 = 89$

   (Mixed numbers)

8. a. $4\frac{5}{11}$ $49 \div 11 = 4$ with a remainder of 5

   b. $8\frac{1}{2}$ $17 \div 2 = 8$ with a remainder of 1

c. $21\frac{4}{9}$ $193 \div 9 = 21$ with a remainder of 4

d. 7 $140 \div 20 = 7$ with no remainder

(Improper fractions)

9. $\dfrac{2}{15}$

$\dfrac{4 \cdot 2}{5 \cdot 2} \cdot \dfrac{3}{4} \cdot \dfrac{2}{3 \cdot 3} = \dfrac{2}{5 \cdot 3} = \dfrac{2}{15}$ Factor all numerators and denominators to create as many common factors as possible. Divide out (cancel) the common factors. It is often not necessary to factor all the way down to primes.

(Multiplication of fractions)

10. $\dfrac{3}{5}$

$\dfrac{4}{3} \cdot \dfrac{3 \cdot 3}{4 \cdot 5} = \dfrac{3}{5}$ **(Multiplication of fractions)**

11. 9

$\dfrac{3}{5} \cdot \dfrac{15}{1} = \dfrac{3}{5} \cdot \dfrac{3 \cdot 5}{1} = \dfrac{3 \cdot 3}{1} = \dfrac{9}{1} = 9$ **(Multiplication of fractions)**

12. $\dfrac{1}{15}$

$\dfrac{9 \cdot 2 \cdot 2 \cdot 2}{5 \cdot 7} \cdot \dfrac{5 \cdot 11}{2 \cdot 9 \cdot 2 \cdot 3} \cdot \dfrac{7}{11 \cdot 2 \cdot 5} = \dfrac{1}{15}$ **(Multiplication of fractions)**

13. $4\frac{4}{5}$

$\dfrac{7 \cdot 3}{5} \cdot \dfrac{8}{7} = \dfrac{24}{5} = 4\frac{4}{5}$ **(Multiplication of fractions)**

14. $18\frac{13}{16}$

$\dfrac{43}{8} \cdot \dfrac{7}{2} = \dfrac{301}{16} = 18\frac{13}{16}$ **(Multiplication of fractions)**

15. $\dfrac{ac^2}{b}$

$\dfrac{a \cdot a \cdot b}{c} \cdot \dfrac{c \cdot c \cdot c}{a \cdot b \cdot b} = \dfrac{a \cdot c^2}{b}$ **(Multiplication of fractions)**

16. $\dfrac{10a}{3c^2}$

$\dfrac{6 \cdot 2 \cdot a \cdot a \cdot b}{c \cdot c \cdot c} \cdot \dfrac{5 \cdot c}{6 \cdot 3 \cdot a \cdot b} = \dfrac{10 \cdot a}{3 \cdot c^2} = \dfrac{10a}{3c^2}$ **(Multiplication of fractions)**

17. $\dfrac{3}{4}$

$\dfrac{5}{18} \cdot \dfrac{27}{10} = \dfrac{5}{9 \cdot 2} \cdot \dfrac{9 \cdot 3}{5 \cdot 2} = \dfrac{3}{4}$ Change division to multiplication by the reciprocal. Factor the numerators and the denominators and divide out (cancel) the common factors. **(Division of fractions)**

18. $\dfrac{15}{16}$

$\dfrac{25}{46} \cdot \dfrac{69}{40} = \dfrac{5 \cdot 5}{23 \cdot 2} \cdot \dfrac{23 \cdot 3}{5 \cdot 8} = \dfrac{15}{16}$ **(Division of fractions)**

19. 128

$\dfrac{16}{1} \cdot \dfrac{8}{1} = \dfrac{128}{1} = 128$ **(Division of fractions)**

20. $\dfrac{3}{10}$

$\dfrac{3}{5} \cdot \dfrac{1}{2} = \dfrac{3}{10}$ **(Division of fractions)**

21. $\dfrac{49}{64}$

$\dfrac{7}{8} \cdot \dfrac{7}{8} = \dfrac{49}{64}$ **(Division of fractions)**

22. $2ab$

$$\frac{12a^2}{5b} \cdot \frac{5b^2}{6a} = \frac{6 \cdot 2 \cdot a \cdot a}{5 \cdot b} \cdot \frac{5 \cdot b \cdot b}{6 \cdot a} = 2ab \quad \textbf{(Division of fractions)}$$

23. $\frac{1}{yx^2}$ or $\frac{1}{x^2y}$

$$\frac{x}{y^2} \cdot \frac{y}{x^3} = \frac{x}{y \cdot y} \cdot \frac{y}{x \cdot x \cdot x} = \frac{1}{y \cdot x^2} = \frac{1}{yx^2} \text{ or } \frac{1}{x^2y} \quad \textbf{(Division of fractions)}$$

24. abc

$$\frac{a^2b}{c^2} \cdot \frac{c^3}{a} = \frac{a \cdot a \cdot b}{c \cdot c} \cdot \frac{c \cdot c \cdot c}{a} = abc \quad \textbf{(Division of fractions)}$$

25. $\frac{5}{23} + \frac{7}{23} + \frac{9}{23}$

Add all numerators and keep the common denominator.

$= \frac{21}{23}$

Reduce the final answer if possible.

(Addition and subtraction of fractions)

26. $\frac{12}{25} - \frac{7}{25}$

Subtract the numerators and keep the common denominator.

$= \frac{5}{25}$

$= \frac{1}{5}$

(Addition and subtraction of fractions)

27. $12\frac{3}{4} + 5\frac{1}{4}$

$= 17\frac{4}{4}$

Add the whole numbers and the fractions.

$= 18$

Simplify the answer.

(Addition and subtraction of fractions)

28. $7\frac{11}{15} - 3\frac{8}{15}$

$= 4\frac{3}{15}$

Subtract the whole numbers and the fractions.

$= 4\frac{1}{5}$

Reduce the final answer.

(Addition and subtraction of fractions)

29. $13\frac{4}{15} + 12\frac{11}{15} + 26\frac{3}{15}$

$= 51\frac{18}{15}$

Change $\frac{18}{15}$ to a mixed number.

$$= 51 + 1 + \frac{3}{15}$$

$$= 52\frac{3}{15}$$

$$= 52\frac{1}{5}$$

(Addition and subtraction of fractions)

30. $12 - 10\frac{11}{16}$

$$= 11\frac{16}{16} - 10\frac{11}{16} \qquad \text{Rewrite 12 as } 11\frac{16}{16}.$$

$$= 1\frac{5}{16}$$

(Addition and subtraction of fractions)

31. $8\frac{1}{6} - 2\frac{5}{6}$

$$= \left(7 + \frac{6}{6} + \frac{1}{6}\right) - 2\frac{5}{6} \qquad \text{Rewrite } 8\frac{1}{6} \text{ as } 7\frac{7}{6}.$$

$$= 7\frac{7}{6} - 2\frac{5}{6}$$

$$= 5\frac{2}{6}$$

$$= 5\frac{1}{3}$$

(Addition and subtraction of fractions)

32. $\frac{9}{24} + \frac{2}{24} + \frac{5}{24}$ Rewrite all as equivalent fractions with a common denominator.

$$= \frac{16}{24} \qquad\qquad\qquad \text{Add all numerators and keep the common denominator.}$$

$$= \frac{2}{3}$$

(Addition and subtraction of fractions)

33. $\dfrac{55}{60} - \dfrac{24}{60}$

    $= \dfrac{31}{60}$

    (Addition and subtraction of fractions)

34. $\dfrac{7}{24} - \dfrac{1}{24}$ Add the first two fractions.

    $= \dfrac{6}{24}$

    $= \dfrac{1}{4}$

    (Addition and subtraction of fractions)

35. $12\dfrac{21}{24} - 3\dfrac{20}{24}$

    $= 9\dfrac{1}{24}$

    (Addition and subtraction of fractions)

36. $\dfrac{35}{80} + \dfrac{12}{80} + \dfrac{16}{80}$

    $= \dfrac{63}{80}$

    (Addition and subtraction of fractions)

37. $\left(11 + \dfrac{10}{10} + \dfrac{3}{10}\right) - 5\dfrac{7}{10}$

    $= 11\dfrac{13}{10} - 5\dfrac{7}{10}$ Rename the first fraction.

    $= 6\dfrac{6}{10}$

    $= 6\dfrac{3}{5}$

    (Addition and subtraction of fractions)

38. $2\dfrac{6}{6} - \dfrac{5}{6}$ Rename the first fraction.

    $= 2\dfrac{1}{6}$

    (Addition and subtraction of fractions)

39. $12\dfrac{22}{24} + 18 + 8\dfrac{15}{24}$

    $= 38\dfrac{37}{24}$

    $= 38 + 1 + \dfrac{13}{24}$

    $= 39\dfrac{13}{24}$

    (Addition and subtraction of fractions)

40. $14\dfrac{2}{3} - 7$

    $= 7\dfrac{2}{3}$

    (Addition and subtraction of fractions)

41. $9\dfrac{8}{12} - 5\dfrac{9}{12}$

    $= \left(8 + \dfrac{12}{12} + \dfrac{8}{12}\right) - 5\dfrac{9}{12}$

    $= 8\dfrac{20}{12} - 5\dfrac{9}{12}$

    $= 3\dfrac{11}{12}$

    (Addition and subtraction of fractions)

42. $\dfrac{95}{210} + \dfrac{39}{210}$

    $= \dfrac{134}{210}$

    $= \dfrac{67}{105}$

    (Addition and subtraction of fractions)

43. $13\dfrac{4}{24} - 12\dfrac{15}{24}$

    $= \left(12 + \dfrac{24}{24} + \dfrac{4}{24}\right) - 12\dfrac{15}{24}$

$= 12\dfrac{28}{24} - 12\dfrac{15}{24}$ Rename the first fraction.

$= \dfrac{13}{24}$

(Addition and subtraction of fractions)

44. $\dfrac{1}{6}$ $\dfrac{1}{8} \div \dfrac{3}{4} = \dfrac{1}{8} \cdot \dfrac{4}{3} = \dfrac{1}{4 \cdot 2} \cdot \dfrac{4}{3} = \dfrac{1}{6}$ **(Simplification of complex fractions)**

45. $2\dfrac{1}{3}$ $\dfrac{2}{3} \div \dfrac{2}{7} = \dfrac{2}{3} \cdot \dfrac{7}{2} = \dfrac{7}{3} = 2\dfrac{1}{3}$ **(Simplification of complex fractions)**

46. 2 $\dfrac{14}{5} \div \dfrac{7}{5} = \dfrac{14}{5} \cdot \dfrac{5}{7} = \dfrac{7 \cdot 2}{5} \cdot \dfrac{5}{7} = \dfrac{2}{1} = 2$

(Simplification of complex fractions)

47. $1\dfrac{1}{22}$ $\dfrac{\frac{15}{20} + \frac{8}{20}}{\frac{5}{10} + \frac{6}{10}} = \dfrac{\frac{23}{20}}{\frac{11}{10}} = \dfrac{23}{20} \div \dfrac{11}{10} = \dfrac{23}{20} \cdot \dfrac{10}{11} = \dfrac{23}{2 \cdot 10} \cdot \dfrac{10}{11} = \dfrac{23}{22} = 1\dfrac{1}{22}$

(Simplification of complex fractions)

48. 3 $\dfrac{\frac{7}{6} + \frac{4}{6}}{\frac{27}{18} - \frac{16}{18}} = \dfrac{\frac{11}{6}}{\frac{11}{18}} = \dfrac{11}{6} \div \dfrac{11}{18} = \dfrac{11}{6} \cdot \dfrac{18}{11} = \dfrac{18}{6} = 3$

(Simplification of complex fractions)

49. $\dfrac{2}{13}$ $\dfrac{\frac{3}{4}}{4\frac{8}{8} - \frac{1}{8}} = \dfrac{\frac{3}{4}}{4\frac{7}{8}} = \dfrac{3}{4} \div \dfrac{39}{8} = \dfrac{3}{4} \cdot \dfrac{8}{39} = \dfrac{3}{4} \cdot \dfrac{4 \cdot 2}{3 \cdot 13} = \dfrac{2}{13}$

(Simplification of complex fractions)

50. $\dfrac{44}{65}$ $\dfrac{\frac{8}{10} + \frac{3}{10}}{\frac{7}{8} + \frac{6}{8}} = \dfrac{\frac{11}{10}}{\frac{13}{8}} = \dfrac{11}{10} \div \dfrac{13}{8} = \dfrac{11}{10} \cdot \dfrac{8}{13} = \dfrac{11}{2 \cdot 5} \cdot \dfrac{2 \cdot 4}{13} = \dfrac{44}{65}$

(Simplification of complex fractions)

51. $3\dfrac{1}{7}$ $\dfrac{2 + \frac{3}{4}}{\frac{8}{8} - \frac{1}{8}} = \dfrac{\frac{11}{4}}{\frac{7}{8}} = \dfrac{11}{4} \div \dfrac{7}{8} = \dfrac{11}{4} \cdot \dfrac{4 \cdot 2}{7} = \dfrac{22}{7} = 3\dfrac{1}{7}$

(Simplification of complex fractions)

52. $\left(\dfrac{7}{8} - \dfrac{4}{8}\right) \div \dfrac{3}{11}$ Do subtraction inside parentheses — LCD needed.

$= \dfrac{3}{8} \div \dfrac{3}{11}$

$$= \frac{3}{8} \cdot \frac{11}{3}$$ Change division to multiplication of the reciprocal.

$$= \frac{11}{8}$$ Divide out common factors.

$$= 1\frac{3}{8}$$ Change to a mixed number. **(Order of operations with fractions)**

53. $$2\frac{3}{8} + \frac{1}{2}\left(\frac{1}{3} + \frac{5}{3}\right)^3$$ Do addition inside parentheses — LCD needed.

$$= 2\frac{3}{8} + \frac{1}{2}\left(\frac{6}{3}\right)^3$$

$$= 2\frac{3}{8} + \frac{1}{2}(8)$$ Evaluate the exponent.

$$= 2\frac{3}{8} + 4$$ Multiply.

$$= 6\frac{3}{8}$$ Add. **(Order of operations with fractions)**

54. $$2\left(\frac{3}{6} + \frac{2}{6}\right) + 3\left(\frac{8}{12} + \frac{3}{12}\right)$$ Do additions in parentheses — LCD needed.

$$2\left(\frac{5}{6}\right) + 3\left(\frac{11}{12}\right)$$

$$= \frac{5}{3} + \frac{11}{4}$$ Do multiplications.

$$= \frac{20}{12} + \frac{33}{12}$$ Do addition — LCD needed.

$$= \frac{53}{12}$$

$$= 4\frac{5}{12}$$ Change to mixed number. **(Order of operations with fractions)**

55. $$\left(\frac{3}{4} \cdot \frac{5}{6}\right) + \left(\frac{3}{4} \cdot \frac{6}{5}\right)$$

$$= \left(\frac{5}{8}\right) + \left(\frac{3}{4} \cdot \frac{6}{5}\right)$$ Divide out common factors and multiply in parentheses.

$$= \frac{5}{8} + \frac{9}{10}$$ Do multiplications in parentheses.

$$= \frac{25}{40} + \frac{36}{40}$$ Do addition — LCD needed.

$= \dfrac{61}{40}$

$= 1\dfrac{21}{40}$ Change to a mixed number. **(Order of operations with fractions)**

56. $(32)\left(\dfrac{9}{16}\right) - \dfrac{1}{2}(48)\left(\dfrac{3}{8}\right)$ Evaluate the exponent.

     $= 18 - 9$ Multiply.

     $= 9$ Subtract. **(Order of operations with fractions)**

57. $\left(\dfrac{3}{4}\right)(3)\left(\dfrac{5}{9}\right) \div \dfrac{25}{64}$ Evaluate the exponent.

     $= \left(\dfrac{9}{4}\right)\left(\dfrac{5}{9}\right) \div \dfrac{25}{64}$ Multiply.

     $= \dfrac{5}{4} \div \dfrac{25}{64}$ Multiply.

     $= \dfrac{5}{4} \cdot \dfrac{64}{25}$ Change division to multiplication by the reciprocal.

     $= \dfrac{16}{5}$ Divide out the common factors (cancel).

     $= 3\dfrac{1}{5}$ Change to a mixed number. **(Order of operations with fractions)**

58. $3 - \dfrac{1}{8}\left[5\dfrac{4}{8} - 4\dfrac{7}{8}\right]^2$ Rewrite the subtraction problem with LCD.

     $= 3 - \dfrac{1}{8}\left[4\dfrac{12}{8} - 4\dfrac{7}{8}\right]^2$ Borrowing required.

     $= 3 - \dfrac{1}{8}\left[\dfrac{5}{8}\right]^2$ Subtract.

     $= 3 - \dfrac{1}{8}\left[\dfrac{25}{64}\right]$ Evaluate the exponent.

     $= 3 - \dfrac{25}{512}$ Multiply.

     $= 2\dfrac{512}{512} - \dfrac{25}{512}$ Borrowing required.

     $= 2\dfrac{487}{512}$ **(Order of operations with fractions)**

59. $3\frac{1}{4} + 1\frac{1}{5}\left[\frac{1}{4} + \frac{8}{4}\right]$ Rewrite with LCD.

$= 3\frac{1}{4} + 1\frac{1}{5}\left[\frac{9}{4}\right]$ Do addition in parentheses.

$= 3\frac{1}{4} + \frac{6}{5}\left[\frac{9}{4}\right]$ Change mixed number to improper fraction.

$= 3\frac{1}{4} + \frac{27}{10}$ Multiply.

$= 3\frac{5}{20} + \frac{54}{20}$ Rewrite with LCD.

$= 3\frac{59}{20}$ Add.

$= 5\frac{19}{20}$ Change $\frac{59}{20}$ into $2\frac{19}{20}$. **(Order of operations with fractions)**

60. $\frac{9}{24} + \left[\frac{1}{8} + \frac{1}{36}\right]$ Evaluate the exponent.

$= \frac{9}{24} + \left[\frac{9}{72} + \frac{2}{72}\right]$ Rewrite with LCD.

$= \frac{9}{24} + \frac{11}{72}$ Add.

$= \frac{27}{72} + \frac{11}{72}$ Rewrite with LCD.

$= \frac{38}{72}$ Add.

$= \frac{19}{36}$ Reduce to lowest terms. **(Order of operations with fractions)**

61. $3 \cdot \frac{1}{3}x = 7 \cdot 3$ Multiply both sides by 3.

$x = 21$ **(Solving equations with fractions)**

62. $\frac{x}{6} + 1 - 1 = \frac{4}{3} - 1$ Subtract 1 from both sides.

$\frac{x}{6} = \frac{4}{3} - \frac{3}{3}$ Change 1 to fraction with LCD.

$\frac{x}{6} = \frac{1}{3}$ Subtract.

$$6 \cdot \frac{x}{6} = \frac{1}{3} \cdot 6$$

Multiply both sides by 6.

$$x = 2$$

(Solving equations with fractions)

63. $$\frac{1}{3}a - \frac{4}{5} + \frac{4}{5} = \frac{8}{15} + \frac{4}{5}$$

Add $\frac{4}{5}$ to both sides.

$$\frac{1}{3}a = \frac{8}{15} + \frac{12}{15}$$

Rewrite with LCD.

$$\frac{1}{3}a = \frac{20}{15}$$

Add.

$$3 \cdot \frac{1}{3}a = \frac{20}{15} \cdot 3$$

Multiply both sides by 3.

$$a = \frac{20}{5}$$

Multiply.

$$a = 4$$

Change to a mixed number. **(Solving equations with fractions)**

64. $$\frac{3}{7} - \frac{3}{7} + \frac{1}{4}c = \frac{13}{28} - \frac{3}{7}$$

Subtract $\frac{3}{7}$ from both sides.

$$\frac{1}{4}c = \frac{13}{28} - \frac{12}{28}$$

Rewrite with LCD.

$$\frac{1}{4}c = \frac{1}{28}$$

Add.

$$4 \cdot \frac{1}{4}c = \frac{1}{28} \cdot 4$$

Multiply both sides by 4.

$$c = \frac{1}{7}$$

Multiply. **(Solving equations with fractions)**

65. $$\frac{y}{5} + 2 - 2 = \frac{11}{5} - 2$$

Subtract 2 from both sides.

$$\frac{y}{5} = \frac{11}{5} - \frac{10}{5}$$

Rewrite with LCD.

$$\frac{y}{5} = \frac{1}{5}$$

Subtract.

$$5 \cdot \frac{y}{5} = \frac{1}{5} \cdot 5$$

Multiply both sides by 5.

$$y = 1$$

Multiply. **(Solving equations with fractions)**

66. $\dfrac{1}{6} - \dfrac{1}{6} + \dfrac{x}{4} = \dfrac{17}{12} - \dfrac{1}{6}$ Subtract $\dfrac{1}{6}$ from both sides.

$\dfrac{x}{4} = \dfrac{17}{12} - \dfrac{2}{12}$ Rewrite with LCD.

$\dfrac{x}{4} = \dfrac{15}{12}$ Subtract.

$4 \cdot \dfrac{x}{4} = \dfrac{15}{12} \cdot 4$ Multiply both sides by 4.

$x = \dfrac{15}{3}$ Multiply.

$x = 5$ Change to a mixed number. **(Solving equations with fractions)**

67. $\dfrac{14}{15} - \dfrac{1}{30} = x + \dfrac{1}{30} - \dfrac{1}{30}$ Subtract $\dfrac{1}{30}$ from both sides.

$\dfrac{28}{30} - \dfrac{1}{30} = x$ Rewrite with LCD.

$\dfrac{27}{30} = x$ Subtract.

$\dfrac{9}{10} = x$ Reduce to lowest terms. **(Solving equations with fractions)**

68. $\dfrac{9}{2} \cdot \dfrac{2}{9} y = 4 \cdot \dfrac{9}{2}$ Multiply both sides by $\dfrac{9}{2}$.

$y = 18$ Multiply. **(Solving equations with fractions)**

69. $\dfrac{3}{7} x - \dfrac{4}{14} + \dfrac{4}{14} = \dfrac{1}{2} + \dfrac{4}{14}$ Add $\dfrac{4}{14}$ to both sides.

$\dfrac{3}{7} x = \dfrac{7}{14} + \dfrac{4}{14}$ Rewrite with LCD.

$\dfrac{3}{7} x = \dfrac{11}{14}$ Add.

$\dfrac{7}{3} \cdot \dfrac{3}{7} x = \dfrac{11}{14} \cdot \dfrac{7}{3}$ Multiply both sides by $\dfrac{7}{3}$.

$x = \dfrac{11}{6}$ Multiply.

$x = 1\dfrac{5}{6}$ Change $\dfrac{11}{6}$ into a mixed number. **(Solving equations with fractions)**

70. $\dfrac{8}{9}y + \dfrac{2}{3} - \dfrac{2}{3} = \dfrac{5}{6} - \dfrac{2}{3}$ Subtract $\dfrac{2}{3}$ from both sides.

$\dfrac{8}{9}y = \dfrac{5}{6} - \dfrac{4}{6}$ Rewrite with LCD.

$\dfrac{8}{9}y = \dfrac{1}{6}$ Subtract.

$\dfrac{9}{8} \cdot \dfrac{8}{9}y = \dfrac{1}{6} \cdot \dfrac{9}{8}$ Multiply both sides by $\dfrac{9}{8}$.

$y = \dfrac{3}{16}$ Multiply. **(Solving equations with fractions)**

71. $17\dfrac{31}{40}$ inches   To find the total, addition must be used.

$11\dfrac{3}{8} + 6\dfrac{2}{5}$

$= 11\dfrac{15}{40} + 6\dfrac{16}{40}$

$= 17\dfrac{31}{40}$ **(Word problems with fractions)**

72. 40 poonie cakes   The word "of" implies multiplication.

$\left(\dfrac{2}{3}\right)(60)$

$= 40$ **(Word problems with fractions)**

73. $2\dfrac{1}{4}$ hours   Subtraction must be used to find the difference between what she said she would practice and what she actually has practiced.

$7 - 4\dfrac{3}{4}$

$= 6\dfrac{4}{4} - 4\dfrac{3}{4}$

$= 2\dfrac{1}{4}$ **(Word problems with fractions)**

74. $23\dfrac{7}{12}$ yards   First find the total amount Linda cut off of the bolt by adding together the two desired lengths. Then subtract the sum from the material initially on the bolt to find out how much is left.

$2\dfrac{3}{4} + 3\dfrac{2}{3}$

$$= 2\frac{9}{12} + 3\frac{8}{12}$$

$$= 5\frac{17}{12}$$

$$= 6\frac{5}{12}$$

$$= 29\frac{12}{12} - 6\frac{5}{12}$$

$$= 23\frac{7}{12} \quad \textbf{(Word problems with fractions)}$$

75. 20 pieces

The 6-inch piece must be divided into pieces $\frac{3}{10}$ inch long.

$$6 \div \frac{3}{10}$$

$$= 6 \cdot \frac{10}{3}$$

$$= 20 \quad \textbf{(Word problems with fractions)}$$

76. 72 laps

The total length of the run must be divided into $\frac{3}{8}$ mile laps.

$$27 \div \frac{3}{8}$$

$$= 27 \cdot \frac{8}{3}$$

$$= 72 \quad \textbf{(Word problems with fractions)}$$

77. $24\frac{5}{8}$ inches

The difference between the total length of the pole and the amount of the pole above the ground will be the amount of the pole under the ground.

$$86\frac{1}{4} - 61\frac{5}{8}$$

$$= 86\frac{2}{8} - 61\frac{5}{8}$$

$$= 85\frac{10}{8} - 61\frac{5}{8}$$

$$= 24\frac{5}{8} \quad \textbf{(Word problems with fractions)}$$

78. $17\dfrac{3}{4}$ yards

The word "of" implies multiplication.

$$\left(\dfrac{3}{4}\right)\left(23\dfrac{2}{3}\right)$$

$$= \left(\dfrac{3}{4}\right)\left(\dfrac{71}{3}\right)$$

$$= \dfrac{71}{4}$$

$$= 17\dfrac{3}{4} \quad \textbf{(Word problems with fractions)}$$

79. $7\dfrac{1}{3}$

Translate into an equation.

$$\dfrac{x}{6} - \dfrac{5}{9} = \dfrac{2}{3}$$

$$\dfrac{x}{6} = \dfrac{2}{3} + \dfrac{5}{9}$$

$$\dfrac{x}{6} = \dfrac{6}{9} + \dfrac{5}{9}$$

$$\dfrac{x}{6} = \dfrac{11}{9}$$

$$6 \cdot \dfrac{x}{6} = \dfrac{11}{9} \cdot 6$$

$$x = \dfrac{22}{3}$$

$$x = 7\dfrac{1}{3} \quad \textbf{(Word problems with fractions)}$$

80. 182 pages

First find the fraction of the book that has already been read by adding.

$$\dfrac{1}{3} + \dfrac{1}{4}$$

$$= \dfrac{4}{12} + \dfrac{3}{12}$$

$$= \dfrac{7}{12} \text{ of the book has been read}$$

The word "of" implies multiplication.

$$\left(\dfrac{7}{12}\right)(312) = 182 \quad \textbf{(Word problems with fractions)}$$

Grade Yourself

Circle the question numbers that you had incorrect. Then indicate the number of questions you missed. If you answered more than three questions incorrectly, you need to focus on that topic. (If a topic has less than three questions and you had at least one wrong, we suggest you study that topic also. Read your textbook or a review book, or ask your teacher for help.)

Subject: Fractions

Topic	Question Numbers	Number Incorrect
Divisibility rules	1	
Prime factorization	2	
Equivalent fractions	3	
Reducing fractions	4, 5, 6	
Mixed numbers	7	
Improper fractions	8	
Multiplication of fractions	9, 10, 11, 12, 13, 14, 15, 16	
Division of fractions	17, 18, 19, 20, 21, 22, 23, 24	
Addition and subtraction of fractions	25, 26, 27, 28, 29, 30, 31, 32, 33, 34, 35, 36, 37, 38, 39, 40, 41, 42, 43	
Simplification of complex fractions	44, 45, 46, 47, 48, 49, 50, 51	
Order of operations with fractions	52, 53, 54, 55, 56, 57, 58, 59, 60	
Solving equations with fractions	61, 62, 63, 64, 65, 66, 67, 68, 69, 70	
Word problems with fractions	71, 72, 73, 74, 75, 76, 77, 78, 79, 80	

Decimals

3

Brief Yourself

This chapter contains a review of the rules for working with decimals. It includes questions and answers about place values, operations with decimals, powers of 10, word problems, exponents, order of operations, and equations. Decimals are another way of writing fractions and mixed numbers.

thousands	hundreds	tens	ones	.	tenths	hundredths	thousandths	ten thousandths
1,000	100	10	1	.	$\frac{1}{10}$	$\frac{1}{100}$	$\frac{1}{1000}$	$\frac{1}{10,000}$
		7	2	.	4	5	7	

IN WORDS   seventy-two and four hundred fifty seven thousandths

IN EXPANDED FORM $(7 \cdot 10) + (2 \cdot 1) + (4 \cdot 1/10) + (5 \cdot 1/100) + (7 \cdot 1/1000)$

                     70   +   2   +   4/10   +   5/100   +   7/1000

When reading or writing a number in decimal notation:

    — read or write the whole number part.

    — the decimal point translates to "and."

    — Read or write the decimal part as though it were a whole number followed by the place value of the digit farthest to the right.

Zeros added to the right of the decimal point, following the last digit, do not change the value of the decimal. They are usually not used.

    $0.3 = 0.30 = 0.300$ These all sound different when read, but all have the same value.

To change a decimal to a fraction:

    — Read the word name.

    — Write the fraction equivalent to that name.

    — Reduce if possible.

    — If there is a whole number part, the result will be a mixed number.

To compare decimal values:

— Make all decimals have the same number of decimal places to the right of the decimal point by adding 0s to the right of each last digit if necessary.

— Ignore the decimal point and compare as if whole numbers.

To round a decimal number to a place value to the right of the decimal point:

— Identify the place value to be rounded.

— Look at the digit immediately to the right of that digit.

— If the digit to the right is 5 or more, the digit in the original place value should be increased by 1 and all other digits to the right of it should be dropped.

— If the digit to the right is less than 5, the digit in the original place value should stay the same and all other digits to the right of it should be dropped.

If the place value being rounded to is to the left of the decimal point, follow the same steps as above, with one exception. The digits between the one that has been rounded and the decimal point should all be changed to zeros. Any digits to the right of the decimal point should still be dropped.

Addition/Subtraction

— Write the numbers so that the decimal points line up vertically.

— Add or subtract as whole numbers.

— Position the decimal point of the answer so that it lines up vertically with the decimal points in the problem.

NOTE:

— Zeros may be added to the right end of any numbers so that it will be easier to keep the number in columns.

— Whole numbers have an understood decimal point at the right end of the number.

— In subtraction, zeros should be added to the top number so that it is possible to borrow.

Example

4.2 – 2.861 becomes

$$\begin{array}{r} 4.200 \\ -2.861 \\ \hline 1.339 \end{array}$$

Multiplication

— Multiply the decimal numbers as if they were whole numbers.

— Place the decimal point in the product so that the number of digits to its right is equal to the total number of digits to the right of the decimal points of the numbers in the original problem.

— Zeros may be added to the left of the product to allow for the correct number of decimal places.

Example

(2.3)(1.82) becomes

$$
\begin{array}{r}
1.82 \\
\times\,2.1 \\
\hline
182 \\
364 \\
\hline
3.822
\end{array}
$$

(2 places to the right)
(1 place to the right)

(3 places to the right)

Division

— Make the divisor a whole number, if necessary, by moving its decimal point to the right end of the number.

— Move the decimal point in the dividend the same number of places and in the same direction as you did with the divisor.

— Zeros can be added if necessary.

— Place the decimal point in the quotient directly over the newly positioned decimal point in the dividend.

— Divide as if whole numbers.

— Round the answer to the given place value.

— If no place value is given, divide until the remainder is 0 or round to an appropriate place value for the problem.

To change a fraction to a decimal:

— Divide the denominator into the numerator.

— Add a decimal point and zeros if necessary.

— If the division does not terminate, round to the desired place value.

— A proper fraction will result in a decimal less than 1.

— With a mixed number, the whole part can be written to the left of the decimal point and the fraction part can be divided out to determine what goes to the right of the decimal point.

Powers of ten are the numbers that result from raising 10 to a power. Because no other digits result from raising 10 to a power except 1's and 0's, some unique things happen when decimal numbers are multiplied or divided by powers of ten. The following are shortcuts that can be used when multiplying or dividing by powers of ten. If the shortcuts are followed, the regular rules can be ignored.

When multiplying a number by powers of 10 such as 10, 100, or 1,000:

— Count the number of zeros in the power.

— Move the decimal point in the original number that number of places to the right.

— Add zeros only if necessary.

When multiplying a number by powers of 10 such as 0.1, 0.01, 0.001:

— Count the total number of decimal place values to the right of the decimal point.

— Move the decimal point in the original number that number of places to the left.

— Add zeros only if necessary.

Examples

(34.5462)(1000) = 34546.2 The decimal point moves three places to the right since 1,000 has 3 zeros.

(34.5462)(0.001) = 0.0345462 The decimal point moves three places to the left since there are three places to the right of the decimal point in this power of 10.

When dividing a number by powers of 10 such as 10, 100, or 1,000:

— Count the number of zeros in the power.

— Move the decimal point in the original number that number of places to the left.

— Add zeros only if necessary.

Examples

761.59 ÷ 100 = 7.6159 The decimal point moves two places to the left since 100 has 2 zeros.

761.59 ÷ (0.001) = 761, 590 The decimal point moves three places to the right since there are three places to the right of the decimal point in this power of 10.

NOTE: These shortcuts are meant to be helpful. However, if they seem to be confusing, remember that all multiplication and division problems can be done by following the usual rules.

Exponents — Combine the rules of exponents and the rules for multiplying decimals. Review the exponent rules in Chapter 1 if necessary.

Order of operations — Combine the rules for order of operations and the rules for decimals. Review the order of operations in Chapter 1 if necessary.

Equations — Combine the rules for equations with the rules of decimals. Review the equation rules in Chapter 1 if necessary.

If a problem contains both fractions and decimals, either all fractions must be changed to decimals or all decimals must be changed to fractions. If a fraction changes to a non-terminating decimal, fractions should be used to evaluate the problem.

 # Test Yourself

1. Write the name of each number in words.

   a. 0.726

   b. 2.04

   c. 5.6

   d. 273.4628

2. Write each of the following as a decimal number.

   a. thirty-three hundredths

   b. four thousand, three hundred two and one hundred thirty-two thousandths

   c. seventy-five and thirty-two thousandths

   d. two hundred and three hundredths

3. Write the following in order from smallest to largest.

   0.02   0.06   0.023   0.056   0.007   0.002

4. Place >, <, or = between the numbers to make each statement true.

   a. 13.1 13.099

b. 7.86 7.68

c. 0.3 0.31

d. 0.26 0.2600

5. Change the following decimals to fractions in lowest terms.

   a. 5.4

   b. 0.027

   c. 0.65

   d. 2.375

6. Round each of the following to the nearest tenth and then to the nearest hundredth.

   a. 0.9392

   b 7.2509

   c. 0.993

   d. 119.5237

Add each of the following.

7. $5.002 + 6.78 + 0.0003$

8. $0.287 + 0.3 + 0.946 + 1.43$

9. $0.083 + 14 + 0.3128 + 6.8$

10. $87.57 + 600.894 + 13.636$

11. $9.495 + 29.147 + 16$

12. $16.3 + 41.08 + 11.029$

Subtract each of the following.

13. $10.565 - 3.865$

14. $62.24 - 9.8$

15. $16 - 3.25$

16. $119.45 - 21.08$

17. $0.32 - 0.0562$

18. $8.27 - 0.396$

Multiply each of the following.

19. $(13.75)(0.43)$

20. $(1.25)(0.08)$

21. $(1.2)(1.2)$

22. $(0.2)(0.3)(0.4)$

23. $(1.4)(0.003)$

24. $(7)(0.07)$

25. $(4.6)(3.42)$

Divide each of the following. Round answers to the nearest hundredth if necessary.

26. $4.425 \div 6$

27. $11.778 \div 1.3$

28. $305 \div 1.4$

29. $1.1982 \div 0.033$

30. $0.144 \div 0.12$

31. $385.32 \div 5.2$

Change each fraction to a decimal. Round to the nearest hundredth if necessary.

32. $\dfrac{3}{8}$

33. $\dfrac{7}{11}$

34. $4\dfrac{2}{9}$

35. $\dfrac{3}{7}$

36. $3\dfrac{5}{16}$

Multiply or divide each of the following using the properties of the powers of 10.

37. $(0.247)(10^5)$

38. $3.53 \div 100$

39. $27.3 \div 0.001$

40. $(1.53)(10000)$

41. $(14.95)(.0001)$

42. $85 \div 1000$

In problems 43 – 46, simplify.

43. $(0.06x)\,(1.2x)$

44. $(7.5t^2)\,(0.3t)$

45. $(0.08x)(3.4x^3)$

46. $(6.4y^2)(0.7y^2)$

In problems 47 – 54, simplify.

47. $12 - 0.2[4.465 - 0.5(0.7 + 1.03)]$

48. $112.6(1.3 - 1.08) - 25.38 \div 2.7$

49. $(6.3)^2 - (15.2 \div 7.6) + 0.2$

50. $2.6 + 0.2(2.6 + 0.2)$

51. $(2.2)^2 + 7.6$

52. $10(0.4)^2 - 0.55$

53. $(5 - 0.06) \div 2(3.42)(0.1)$

54. $10^2 \{ [(3 - 0.24) \div 2.4] - (0.21 - 0.092) \}$

Solve each of the following equations.

55. $x + 3.3 = 43.73$

56. $y - 5 = 18.97$

57. $\dfrac{x}{24} = 0.48$

58. $0.7x - 22.6 = 16.6$

59. $35.755 = 2.8x - 79.045$

60. $4x + 4.78 = 7.5$

61. A can of beer contains 9.2 grams of carbohydrates. How many grams of carbohydrates are there in 4 cases of beer if each case has 24 cans in it?

62. If a glass holds 6.8 ounces, how many glasses can be poured from a jug that holds 136 ounces?

63. Tom buys four cans of tuna costing $1.09 each, three pounds of hamburger at $1.79 per pound, and a magazine for $2.25. If he has a $20 bill, how much change will he receive?

64. If the phone company charges $.53 for the first minute and $.27 for each additional minute, how much will a 15-minute telephone call cost?

65. Sue earns $6.32 per hour for each of the first 36 hours she works in a week. She earns $9.50 per hour in overtime pay for each additional hour she works in the same week. How much money will Sue earn if she works 52 hours in one week?

66. Allan has $352.26 in his checking account. If he writes checks for $25.02, $32.98, and $145, how much money will be left in his account?

67. Mike is fertilizing his garden. He uses 5.6 ounces of fertilizer per square yard. The garden measures 60.5 square yards. How many ounces of fertilizer does Mike need for his garden?

68. It costs $24.95 a day plus $.27 per mile to rent a compact car. What would be the total cost if a compact car is rented for three days and driven 250 miles?

69. On a six-day trip, a driver bought the following amounts of gasoline: 23.6 gallons, 17.7 gallons, 20.8 gallons, 17.2 gallons, 25.4 gallons, 13.8 gallons. How many gallons of gasoline were purchased?

70. How much income does a state collect from the sale of 100,000 lottery tickets at $1.50 each?

In problems 71 – 77, simplify.

71. $\dfrac{19}{50}(1.32 + 0.48)$

72. $\dfrac{1}{2} + 0.75\left(\dfrac{2}{5}\right)$

73. $\dfrac{3}{4}(1.8 + 7.6)$

74. $(6)\left(\dfrac{3}{5}\right)(0.02)$

75. $\left(\dfrac{7}{8}\right) + 0.45\left(\dfrac{3}{4}\right)$

76. $(0.25)^2 + \left(\dfrac{1}{4}\right)^2(3)$

77. $3.4 - \left(\dfrac{1}{2}\right)(0.76)$

Check Yourself

1. a. seven hundred twenty-six thousandths

   b. two and four hundredths

   c. five and six tenths

   d. two hundred seventy-three and four thousand six hundred twenty-eight ten thousandths

   (Introduction to decimals)

2. a. 0.33

   b. 4,302.132

   c. 75.032

   d. 200.03

   (Introduction to decimals)

3.   0.002 0.007 0.02 0.023 0.056 0.06   Compare digits in the same place values. **(Introduction to decimals)**

4. a. $13.1 > 13.099$ Add two zeros to the right of 13.1, ignore decimal points, and compare: $13100 > 13099$.

   b. $7.86 > 7.68$ Ignore decimal points and compare: $786 > 768$.

   c. $0.3 < 0.31$ Add one zero to the right of 0.3, ignore decimal points, and compare: $30 < 31$.

   d. $0.26 = 0.2600$ Add two zeros to the right of 0.26, ignore decimal points, and compare: $2600 = 2600$.

   (Introduction to decimals)

5. a. $5\frac{2}{5}$ It is read $5\frac{4}{10}$. Four tenths reduces to two fifths.

   b. $\frac{27}{1000}$ The decimal has three places to the right of the decimal point, which means there should be three zeros in the denominator.

   c. $\frac{13}{20}$ It is read $\frac{65}{100}$. This reduces to $\frac{13}{20}$.

   d. $2\frac{3}{8}$ It is read $2\frac{375}{1000}$. $\frac{375}{1000}$ reduces to $\frac{3}{8}$.

   (Changing decimals to fractions)

	Tenth	Hundredth
6. a.	0.9	0.94
b.	7.3	7.25
c.	1.0	0.99
d.	119.5	119.52

(Rounding decimals)

7.
```
  5.002
  6.78
  0.0003
 11.7823
```
(Addition of decimals)

8.
```
0.287
0.3
0.946
1.43
2.963
```
(Addition of decimals)

9.
```
  0.083
14.
  0.3128
  6.8
 21.1958
```
The decimal point in a whole number is at the right end of the number. **(Addition of decimals)**

10.
```
  87.57
 600.894
  13.636
 702.100
```
(Addition of decimals)

11.
```
  9.495
 29.147
 16.000
 54.642
```
(Addition of decimals)

12.
```
16.3
41.08
11.029
68.409
```
(Addition of decimals)

13.
```
10.565
 3.865
 6.7
```
(Subtraction of decimals)

14.
```
62.24
 9.8
52.44
```
(Subtraction of decimals)

15.
```
16.00
 3.25
12.75
```
(Subtraction of decimals)

16. 119.45 **(Subtraction of decimals)**
      21.08
     ―――――
      98.37

17. 0.3200 **(Subtraction of decimals)**
     0.0562
     ―――――
     0.2638

18. 8.270 **(Subtraction of decimals)**
     0.396
     ―――――
     7.874

19. 13.75 (2 places to the right) **(Multiplication of decimals)**
       .43 (2 places to the right)
     ―――――
      4125
     5 500
     ―――――
     5.9125 (4 places to the right)

20. 1.25 (2 places to the right) **(Multiplication of decimals)**
      .08 (2 places to the right)
     ―――――
     1000
      000
     ―――――
     .1000 (4 places to the right)

21. 1.2 (2 places to the right) **(Multiplication of decimals)**
     1.2 (2 places to the right)
     ―――
      24
     12
     ―――
     1.44 (4 places to the right)

22. 0.2 (1 place to the right)   0.06 (2 places to the right)
     0.3 (1 place to the right)   0.4 (1 place to the right)
     ―――   ――――
      06   0.024 (3 places to the right)
     00
     ―――
     0.06 (2 places to the right)

(Multiplication of decimals)

23. 0.003   It is easier to multiply when the longer number is on top. Multiplication is commutative,
       1.4   so the order doesn't matter.
     ―――――
      012
      003
     ―――――
     0.0042

(Multiplication of decimals)

24. 0.07 **(Multiplication of decimals)**
        7
     ―――
     0.49

25. 3.42   **(Multiplication of decimals)**
      4.6
     ―――――
     2 052
     13 68
     ―――――
     15.732 (3 places to the right)

26.
```
       .7375
   6)4.4250
     4 2
       22
       18
       45
       42
        30
        30
         0
```
It is not necessary to shift any decimal points since 6 is already a whole number.

The decimal point comes straight up, and you can divide as with whole numbers.

(Division of decimals)

27.
```
        9.06
   13)117.78
      117
        07
         0
        78
        78
         0
```
This shows the decimal points of both the divisor and the dividend already shifted one place to the right.

(Division of decimals)

28.
```
       217.857
   14)3050.000
      28
       25
       14
      110
       98
      120
      112
        80
        70
       100
        98
         2
```
Rounded to the nearest hundredth, the answer is 217.86.

This shows the decimal points in the divisor and the dividend already shifted 1 place to the right. A zero needed to be appended to the 305 to make the shift possible.

(Division of decimals)

29.
```
        36.309
   33)1198.200
      99
      208
      198
      102
       99
       30
        0
      300
      297
        3
```
Rounded to the nearest hundredth, the answer is 36.31.

This shows the decimal points already shifted 3 places to the right.

(Division of decimals)

30.

$$
\begin{array}{r}
1.2 \\
12\overline{)14.4} \\
\underline{12} \\
24 \\
\underline{24} \\
0
\end{array}
$$

This shows the decimal points already shifted 2 places to the right.

(Division of decimals)

31.

$$
\begin{array}{r}
74.1 \\
52\overline{)3853.2} \\
\underline{364} \\
213 \\
\underline{208} \\
52 \\
\underline{52} \\
0
\end{array}
$$

(Division of decimals)

32. 0.375

$$
\begin{array}{r}
.375 \\
8\overline{)3.000}
\end{array}
$$

(Changing fractions to decimals)

33. 0.64

$$
\begin{array}{r}
0.636 \\
11\overline{)7.000}
\end{array}
$$

(Changing fractions to decimals)

34. 4.22

$$
\begin{array}{r}
0.222 \\
9\overline{)2.000}
\end{array}
$$

The whole number remains the same, so it does not have to be part of the division. If the mixed number is changed to an improper fraction first, the answer will, of course, be the same. **(Changing fractions to decimals)**

35. 0.43

$$
\begin{array}{r}
0.428 \\
7\overline{)3.000}
\end{array}
$$

(Changing fractions to decimals)

36. 3.31

$$
\begin{array}{r}
.312 \\
5\overline{)5.000}
\end{array}
$$

(Changing fractions to decimals)

37. 24,700

10^5 has 5 zeros, so the decimal point shifts 5 places to the right. **(Multiplication by powers of 10)**

38. .0353

100 has 2 zeros, so the decimal point shifts two places to the left. **(Division by powers of 10)**

39. 27,300

0.001 has 3 places to the right of the decimal point, so the decimal point shifts 3 places to the right. **(Division by powers of 10)**

40. 15,300

10,000 has 4 zeros, so the decimal point shifts 4 places to the right. **(Multiplication by powers of 10)**

41. .001495

0.0001 has 4 places to the right of the decimal point, so the decimal point shifts 4 places to the left. **(Division by powers of 10)**

42. .085   1,000 has 3 zeros, so the decimal point shifts 3 places to the left. **(Division by powers of 10)**

43. $.072x^2$ Multiply the numbers using decimal rules for the placement of the decimal point. When multiplying like bases, leave the bases the same and add the exponents. Variables without an exponent written are assumed to have an exponent of 1. **(Multiplication of decimals)**

44. $2.25t^3$ **(Multiplication of decimals)**

45. $.272x^4$ **(Multiplication of decimals)**

46. $4.48y^4$ **(Multiplication of decimals)**

47. $12 - 0.2[4.465 - 0.5(1.73)]$

     $= 12 - 0.2[4.465 - 0.865]$ Multiply within the bracket.

     $= 12 - 0.2[3.6]$ Subtract within the bracket.

     $= 12 - 0.72$ Multiply before subtraction.

     $= 11.28$ Subtract. **(Order of operations using decimals)**

48. $112.6(1.3 - 1.08) - 25.38 \div 2.7$

     $= 112.60(0.22) - 25.38 \div 2.7$ Subtract inside the parentheses.

     $= 24.772 - 25.38 \div 2.7$ Multiply before dividing, going from left to right.

     $= 24.772 - 9.4$ Divide before addition or subtraction.

     $= 15.372$ Subtract. **(Order of operations using decimals)**

49. $(6.3)^2 - (15.2 \div 7.6) + 0.2$

     $= (6.3)^2 - (2) + 0.2$ Divide inside the parentheses.

     $= 39.69 - 2 + 0.2$ Evaluate exponents.

     $= 37.69 + 0.2$ Subtract before addition, going from left to right

     $= 37.89$ Add. **(Order of operations using decimals)**

50. $2.6 + 0.2(2.6 + 0.2)$

     $= 2.6 + 0.2(2.8)$ Add inside the parentheses.

     $= 2.6 + 0.56$ Multiply before addition.

     $= 3.16$ Add. **(Order of operations using decimals)**

51. $(2.2)^2 + 7.6$

     $4.84 + 7.6$ Evaluate the exponent.

     12.44 Add. **(Order of operations using decimals)**

52. $10\,(0.4)^2 - 0.55$

     $= 10(.16) - 0.55$ Evaluate the exponent.

     $= 1.6 - 0.55$ Multiply.

     $= 1.05$ Subtract. **(Order of operations using decimals)**

53. $(5 - 0.06) \div 2\,(3.42)\,(0.1)$

     $= 4.94 \div 2\,(3.42)\,(0.1)$ Subtract inside the parentheses.

     $= (2.47)\,(3.42)\,(0.1)$ Divide before multiplying, going from left to right.

     $= (8.4474)\,(0.1)$ Multiply.

     $= .84474$ Multiply. **(Order of operations using decimals)**

54. $10^2\,\{\,[\,(3 - 0.24) \div 2.4\,] - (0.21 - 0.092)\,\}$

     $= 10^2\,\{\,[2.76 \div 2.4\,] - (0.21 - 0.092)\,\}$ Subtract inside parentheses.

     $= 10^2\,\{1.15 - (0.21 - 0.092)\,\}$ Divide inside brackets.

     $= 10^2\,\{1.15 - 0.118\}$ Subtract inside parentheses.

     $= 10^2\,\{1.032\}$ Subtract inside braces.

     $= 100\,\{1.032\}$ Evaluate the exponent.

     $= 103.2$ **(Order of operations using decimals)**

55. $x + 3.3 - 3.3 = 43.73 - 3.3$ Isolate the variable by subtracting 3.3 from both sides.

              $x = 40.43$ **(Solving equations with decimals)**

56. $y - 5 + 5 = 18.97 + 5$ Add 5 to both sides.

          $y = 23.97$ Remember that the 5 is really 5., and the decimal points need to be aligned for addition. **(Solving equations with decimals)**

57. $24 \cdot \dfrac{x}{24} = (0.48)\,(24)$ Multiply both sides by 24.

         $x = 11.52$ **(Solving equations with decimals)**

58. $0.7x - 22.6 + 22.6 = 16.6 + 22.6$ Add 22.6 to both sides.

             $0.7x = 39.2$

          $\dfrac{0.7x}{0.7} = \dfrac{39.2}{0.7}$ Divide both sides by 0.7.

               $x = 56$ **(Solving equations with decimals)**

59. $35.755 + 79.045 = 2.8x - 79.045 + 79.045$ Add 79.045 to both sides.

$114.8 = 2.8x$

$$\frac{114.8}{2.8} = \frac{2.8x}{2.8}$$ Divide both sides by 2.8.

$41 = x$ **(Solving equations with decimals)**

60. $4x + 4.78 - 4.78 = 7.5 - 4.78$ Subtract 4.78 from both sides.

$4x = 2.72$

$$\frac{4x}{4} = \frac{2.72}{4}$$ Divide both sides by 4.

$x = 0.68$ **(Solving equations with decimals)**

61.   883.2 grams — Multiply 24 cans per case by 4 cases to get the total number of cans (96). Then multiply the number of cans by the number of carbohydrates per can (96)(9.2). **(Word problems using decimals)**

62.   20 glasses — Divide when a larger amount needs to be broken into smaller amounts $(136 \div 6.8)$. **(Word problems using decimals)**

63. $8.02 in change — This problem requires multiplication, addition, and subtraction. To find the total cost of the tuna, multiply $(1.09)(4) = 4.36$. To find the total cost of the hamburger, multiply $(1.79)(3) = 5.37$. To find the cost of all three items, add $4.36 + 5.37 + 2.25 = 11.98$. To find the change, subtract $20.00 - 11.98 = 8.02$. **(Word problems using decimals)**

64. $4.31 — The charge for the 14 additional minutes is found by multiplying $(14)(\$.27) = 3.78$. This gets added to the $.53 charge for the first minute, making the total $4.31. **(Word problems using decimals)**

65. $379.52 — To find the total amount earned for the first 36 hours multiply $(\$6.32)(36) = \$227.52$. To find the amount paid in overtime, multiply $(\$9.50)(16) = \$152.00$. To find the total earned, add the two amounts together. **(Word problems using decimals)**

66. $149.26 — Add to find the total of all checks written. Then subtract this amount from the balance in the account. **(Word problems using decimals)**

67.   338.8 ounces — To find the total amount of fertilizer needed, multiply the number of square yards to be covered times the amount of fertilizer needed to cover each square yard. $(60.5)(5.6) = 338.8$. **(Word problems using decimals)**

68. $142.35 — The total rental fee for the car would be $(24.95)(3) = 74.85$. The total amount charged for the mileage would be $(0.27)(250) = 67.50$. This would make the total charge $74.85 + 67.50 = \$142.35$. **(Word problems using decimals)**

69.   118.5 gallons — To find the total amount purchased, add all of the purchases together $23.6 + 17.7 + 20.8 + 17.2 + 25.4 + 13.8 = 118.5$. **(Word problems using decimals)**

70. $150,000

To find the total amount earned, multiply the income from each ticket times the number of tickets sold.

$(1.50)(100,000) = 150,000.$ **(Word problems using decimals)**

71. $\dfrac{171}{250}$ or .684

$\dfrac{19}{50}\left(1\dfrac{32}{100} + \dfrac{48}{100}\right)$

$(.38)(1.8)$

$= \dfrac{19}{50}\left(1\dfrac{80}{100}\right)$

.684

$= \dfrac{19}{50}\left(\dfrac{180}{100}\right)$

$= \dfrac{171}{250}$ **(Operations with fractions and decimals)**

72. $\dfrac{4}{5}$ or .8

$\dfrac{1}{2} + \left(\dfrac{3}{4}\right)\left(\dfrac{2}{5}\right)$

$0.5 + (0.75)\,(0.4)$

$= \dfrac{1}{2} + \dfrac{3}{10}$

$= 0.5 + 0.3$

$= \dfrac{5}{10} + \dfrac{3}{10}$

$= 0.8$

$= \dfrac{8}{10}$

$= \dfrac{4}{5}$ **(Operations with fractions and decimals)**

73. $7\dfrac{1}{20}$ or 7.05

$\dfrac{3}{4}\left(1\dfrac{8}{10} + 7\dfrac{6}{10}\right)$

$0.75(1.8 + 7.6)$

$= \dfrac{3}{4}\left(8\dfrac{14}{10}\right)$

$= 0.75(9.4)$

$= \left(\dfrac{3}{4}\right)\left(\dfrac{94}{10}\right)$

$= 7.05$

$= \dfrac{141}{20}$

$= 7\dfrac{1}{20}$ **(Operations with fractions and decimals)**

74. $\dfrac{9}{125}$ or 0.072

$\left(\dfrac{6}{1}\right)\left(\dfrac{3}{5}\right)\left(\dfrac{2}{100}\right)$

$(6)(0.6)(0.02)$

$= \dfrac{9}{125}$

$= .072$

(Operations with fractions and decimals)

75. $1\dfrac{17}{80}$ or 1.2125

$\dfrac{7}{8} + \left(\dfrac{45}{100}\right)\left(\dfrac{3}{4}\right)$

$0.875 + (0.45)(0.75)$

$= \dfrac{7}{8} + \dfrac{135}{400}$

$= 0.875 + 0.3375$

$= \dfrac{350}{400} + \dfrac{135}{400}$

$= 1.2125$

$= \dfrac{485}{400}$

$$= 1\frac{85}{400}$$

$$= 1\frac{17}{80} \text{ (Operations with fractions and decimals)}$$

76. $\frac{1}{4}$ or 0.25

$$\left(\frac{1}{4}\right)^2 + \left(\frac{1}{4}\right)^2 (3) \qquad\qquad (0.25)^2 + (0.25)^2 (3)$$

$$= \left(\frac{1}{16}\right) + \left(\frac{1}{16}\right)(3) \qquad = (0.0625) + (0.0625)(3)$$

$$= \frac{1}{16} + \frac{3}{16} \qquad\qquad = 0.0625 + 0.1875$$

$$= \frac{4}{16} \qquad\qquad\qquad = 0.25$$

$$= \frac{1}{4} \text{ (Operations with fractions and decimals)}$$

77. $3\frac{1}{50}$ or 3.02

$$3\frac{4}{10} - \left(\frac{1}{2}\right)\left(\frac{76}{100}\right) \qquad 3.4 - (0.5)(0.76)$$

$$= 3\frac{4}{10} - \frac{38}{100} \qquad\qquad = 3.4 - 0.38$$

$$= 3\frac{40}{100} - \frac{38}{100} \qquad\qquad = 3.02$$

$$= 3\frac{2}{100}$$

$$= 3\frac{1}{50} \text{ (Operations with fractions and decimals)}$$

Grade Yourself

Circle the question numbers that you had incorrect. Then indicate the number of questions you missed. If you answered more than three questions incorrectly, you need to focus on that topic. (If a topic has less than three questions and you had at least one wrong, we suggest you study that topic also. Read your textbook or a review book, or ask your teacher for help.)

Subject: Decimals

Topic	Question Numbers	Number Incorrect
Introduction to decimals	1, 2, 3, 4	
Changing decimals to fractions	5	
Rounding decimals	6	
Addition of decimals	7, 8, 9, 10, 11, 12	
Subtraction of decimals	13, 14, 15, 16, 17, 18	
Multiplication of decimals	19, 20, 21, 22, 23, 24, 25, 43, 44, 45, 46	
Division of decimals	26, 27, 28, 29, 30, 31	
Changing fractions to decimals	32, 33, 34, 35, 36	
Multiplication by powers of 10	37, 40, 41	
Division by powers of 10	38, 39, 42	
Order of operations using decimals	47, 48, 49, 50, 51, 52, 53, 54	
Solving equations with decimals	55, 56, 57, 58, 59, 60	
Word problems using decimals	61, 62, 63, 64, 65, 66, 67, 68, 69, 70	
Operations with fractions and decimals	71, 72, 73, 74, 75, 76, 77	

Ratio and Proportion

4

Brief Yourself

This chapter contains a review of the rules of working with ratios, rates, and proportions. The questions and answers in this chapter will incorporate the rules for the operations of whole numbers, fractions, and decimals with the rules for ratios, rates, and proportions.

Ratio is a comparison of two quantities that have the same units.

Ratios can be written

— Using the word "to"   a to b

— Using a colon instead of "to"   a:b

— As a fraction where the fraction bar replaces "to" $\frac{a}{b}$ (when written as a fraction, should be reduced to

lowest terms)

— Without units

Rate is a comparison of two quantities that are measured in different units with the property that neither unit can be converted into the other. Rates are usually written in fraction form.

Unit rate is a rate that has been simplified so that the numerical part of the denominator is equal to 1.

Example

$$\frac{125 \text{ miles}}{2 \text{ hours}}$$ This is a rate.

$$\frac{62.5 \text{ miles}}{1 \text{ hour}}$$ This is a unit rate. A common way to express this is

62.5 mph.

It is sometimes possible to change different units into the same units. For example,

$$\frac{5 \text{ nickels}}{3 \text{ dimes}}$$ could be written as $\frac{5 \text{ nickels}}{6 \text{ nickels}}$.

Proportion is a mathematical statement in which two ratios or two rates are set equal to each other. The four numbers of a proportion are called terms.

Example

$$\frac{a}{b} = \frac{c}{d}$$

a is the first term   c is the third term
b is the second term   d is the fourth term

The first and fourth terms are called the extremes; the second and the third terms are called the means.

In all true proportions, the product of the means is equal to the product of the extremes.

Example

If $\frac{a}{b} = \frac{c}{d}$, then (a)(d) = (b)(c). Sometimes this is called "cross multiplying."

When one of the terms of the proportion is missing, it is necessary to "solve" the proportion. That is, the value of the variable that makes the proportion true must be found. To do this, we find the value of the variable that makes the cross products equal.

To solve a proportion:

— Multiply the means.

— Multiply the extremes.

— Set the products equal to each other.

— Solve the resulting equation.

NOTE: The first and second step can be interchanged, and the results will not be affected.

When solving a word problem that uses proportions, it is important to make sure that the units in the numerators match and the units in the denominators match.

Example

$$\frac{2 \text{ pounds}}{\$48} = \frac{12 \text{ pounds}}{\$X}$$

$$(2)\,(x) = (12)\,(48)$$

$$2x = 576$$

$$x = \$288$$

Test Yourself

In problems 1 – 11, write each ratio as a fraction in simplest form.

1. 5 to 4

2. 10 to 15

3. 8 to 12

4. 90 to 75

5. 24 to 12

6. $\dfrac{6}{5}$ to $\dfrac{6}{7}$

7. 0.3 to 3

8. 6.4 to 0.8

9. $2\dfrac{2}{3}$ to $\dfrac{5}{3}$

10. $\dfrac{1}{4}$ to 0.75

11. $\dfrac{9}{5}$ to $\dfrac{11}{5}$

In problems 12 – 16, write the rates as fractions in simplest form.

12. 16 feet to 6 seconds

13. 8.6 milligrams to 8 hours

14. 85 miles to 15 hours

15. 12 commercials in 45 minutes

16. 6.5 bushels from 10 trees

Write each of the following as a unit rate.

17. 16 feet to 6 seconds

18. 8.6 milligrams to 8 hours

19. 85 miles in 15 hours

20. 12 commercials in 45 minutes

21. 6.5 bushels from 10 trees

Determine if each of the following proportions is true or false.

22. $\dfrac{7}{5} = \dfrac{84}{60}$

23. $\dfrac{7}{8} = \dfrac{11}{12}$

24. $\dfrac{20}{11} = \dfrac{187}{103}$

25. $\dfrac{3\frac{1}{2}}{2\frac{1}{3}} = \dfrac{9}{6}$

26. $\dfrac{16.24}{24.08} = \dfrac{406}{602}$

27. $\dfrac{2.1}{3.2} = \dfrac{1.2}{2.3}$

28. $\dfrac{0.3}{1.2} = \dfrac{1}{4}$

29. $\dfrac{2}{\frac{1}{4}} = \dfrac{4}{\frac{1}{2}}$

30. $\dfrac{5}{2\frac{2}{3}} = \dfrac{2}{1\frac{1}{15}}$

In problems 31 through 43, solve the proportions.

31. $\dfrac{4}{5} = \dfrac{x}{3}$

32. $\dfrac{x}{10} = \dfrac{40}{25}$

33. $\dfrac{20}{24} = \dfrac{x}{18}$

34. $\dfrac{6}{x} = \dfrac{15}{55}$

35. $\dfrac{10}{\frac{1}{5}} = \dfrac{150}{x}$

36. $\dfrac{4}{x} = \dfrac{\frac{2}{5}}{10}$

37. $\dfrac{0.5}{x} = \dfrac{1.25}{5}$

38. $\dfrac{x}{12} = \dfrac{5}{0.6}$

39. $\dfrac{x}{1.5} = \dfrac{2.5}{7.5}$

40. $\dfrac{\frac{3}{5}}{\frac{3}{8}} = \dfrac{x}{10}$

41. $\dfrac{2.5}{\frac{1}{2}} = \dfrac{\frac{1}{3}}{x}$

42. $\dfrac{8}{x} = \dfrac{3\frac{1}{2}}{10\frac{1}{2}}$

43. $\dfrac{12}{0.23} = \dfrac{9}{x}$

44. Chuck can read six pages in 24 minutes. How many pages can he read in 60 minutes?

45. A recipe calls for four cups of flour and three tablespoons of sugar. How much sugar is needed if the recipe is increased so that it will use 28 cups of flour?

46. A machine can produce 15 engine blocks in two days. At this rate, how many engine blocks can be made in an 8-day period?

47. The scale on the map indicates that $\frac{1}{4}$ inch represents three miles. What is the actual distance between cities that are five inches apart?

48. A car travels 165 miles in three hours. If it continues to travel at this same rate, how far will it travel in eight hours?

49. The ratio of women to men at the college is 7 to 5. How many male students are there if there are 3,500 women?

50. If Joan invests $2,000 in a stock that pays a one-year dividend of $240, how much does she need to invest if she wants to make $600?

51. A metal bar expands $\frac{1}{4}$ inch for every 12° rise in temperature. How much will it expand if the temperature rises 48°?

52. A $3\frac{1}{2}$ pound roast costs $14. At this same price per pound, what size roast can be purchased for $22?

53. If your car can go 120 miles on $3\frac{1}{2}$ gallons of gasoline, how much gasoline will be necessary to go on a 300-mile trip?

54. A biologist tags and releases 500 fish. Later he nets 200 fish and finds only 4 tagged. Estimate the population of fish, according to his sample.

Check Yourself

1. $\dfrac{5}{4}$ The first number becomes the numerator; the second number becomes the denominator. (**Write as a simplified ratio**)

2. $\dfrac{10}{15} = \dfrac{2}{3}$ (**Write as a simplified ratio**)

3. $\dfrac{8}{12} = \dfrac{2}{3}$ (**Write as a simplified ratio**)

4. $\dfrac{90}{75} = \dfrac{6}{5}$ (**Write as a simplified ratio**)

5. $\dfrac{24}{12} = \dfrac{2}{1}$ (**Write as a simplified ratio**)

6. $\dfrac{\frac{6}{5}}{\frac{6}{7}} = \dfrac{6}{5} \div \dfrac{6}{7} = \dfrac{6}{5} \cdot \dfrac{7}{6} = \dfrac{7}{5}$ **(Write as a simplified ratio)**

7. $\dfrac{0.3}{3} = \dfrac{3}{30}$ Multiply both the numerator and the denominator by 10 to eliminate the decimal.

      $= \dfrac{1}{10}$ Reduce to lowest terms. **(Write as a simplified ratio)**

8. $\dfrac{6.4}{0.8} = \dfrac{64}{8} = \dfrac{8}{1}$ **(Write as a simplified ratio)**

9. $\dfrac{\frac{8}{3}}{\frac{5}{3}} = \dfrac{8}{3} \div \dfrac{5}{3} = \dfrac{8}{3} \cdot \dfrac{3}{5} = \dfrac{8}{5}$ Change the mixed number to an improper fraction first.

                             (Write as a simplified ratio)

10. $\dfrac{\frac{1}{4}}{0.75} = \dfrac{\frac{1}{4}}{\frac{3}{4}} = \dfrac{1}{4} \div \dfrac{3}{4} = \dfrac{1}{4} \cdot \dfrac{4}{3} = \dfrac{1}{3}$ Change 0.75 into the fraction $\dfrac{3}{4}$.

      OR

      $\dfrac{0.25}{0.75} = \dfrac{25}{75} = \dfrac{1}{3}$ Change $\dfrac{1}{4}$ into the decimal 0.25. Shift decimal points by multiplying by 100. **(Write as a simplified ratio)**

11. $\dfrac{\frac{9}{5}}{\frac{11}{5}} = \dfrac{9}{5} \div \dfrac{11}{5} = \dfrac{9}{5} \cdot \dfrac{5}{11} = \dfrac{9}{11}$ **(Write as a simplified ratio)**

12. $\dfrac{16 \text{ feet}}{6 \text{ seconds}}$

      $\dfrac{8 \text{ feet}}{3 \text{ seconds}}$ Reduce to lowest terms. **(Write as a simplified rate)**

13. $\dfrac{8.6 \text{ milligrams}}{8 \text{ hours}}$ Multiply the numerator and denominator by 10 to shift the decimal point.

      $\dfrac{86 \text{ milligrams}}{80 \text{ hours}}$

      $\dfrac{43 \text{ milligrams}}{40 \text{ hours}}$ Reduce to lowest terms. **(Write as a simplified rate)**

14. $\dfrac{85 \text{ miles}}{15 \text{ hours}}$

   $\dfrac{17 \text{ miles}}{3 \text{ hours}}$ Reduce to lowest terms. **(Write as a simplified rate)**

15. $\dfrac{12 \text{ commercials}}{45 \text{ minutes}}$

   $\dfrac{4 \text{ commercials}}{15 \text{ minutes}}$ Reduce to lowest terms. **(Write as a simplified rate)**

16. $\dfrac{6.5 \text{ bushels}}{10 \text{ trees}}$

   $\dfrac{65 \text{ bushels}}{100 \text{ trees}}$ Multiply numerator and denominator by 10 to shift the decimal.

   $\dfrac{13 \text{ bushels}}{20 \text{ trees}}$ Reduce to lowest terms. **(Write as a simplified rate)**

17. $\dfrac{16 \text{ feet}}{6 \text{ seconds}}$ Divide 6 into 16.

   $\dfrac{2.67 \text{ feet}}{\text{second}}$ This answer has been rounded to the nearest hundredth.

   $\dfrac{2\frac{2}{3} \text{ feet}}{\text{second}}$ This is the answer in fraction form. **(Write as a unit rate)**

18. $\dfrac{8.6 \text{ milligrams}}{8 \text{ minutes}}$ Divide 8 into 8.6.

   $\dfrac{1.075 \text{ milligrams}}{\text{minute}}$ **(Write as a unit rate)**

19. $\dfrac{85 \text{ miles}}{15 \text{ hours}}$ Divide 15 into 85.

   $\dfrac{5.67 \text{ miles}}{\text{hour}}$ This answer has been rounded to the nearest hundredth.

   $\dfrac{5\frac{2}{3} \text{ miles}}{\text{hour}}$ This is the answer in fraction form. **(Write as a unit rate)**

20. $\dfrac{12 \text{ commercials}}{45 \text{ minutes}}$ Divide 45 into 12.

$\dfrac{0.27 \text{ commecials}}{\text{minute}}$ This answer has been rounded to the nearest hundredth.

$\dfrac{\frac{4}{15} \text{ commercials}}{\text{minute}}$ This is the answer in fraction form. **(Write as a unit rate)**

21. $\dfrac{6.5 \text{ bushels}}{10 \text{ trees}}$ Divide 10 into 6.5.

$\dfrac{0.65 \text{ bushels}}{\text{tree}}$

$\dfrac{\frac{13}{20} \text{ bushels}}{\text{tree}}$ This is the answer in fraction form. **(Write as a unit rate)**

22. True

$(7)(60) = (5)(84)$ Cross multiply.

$420 = 420$ The cross products are equal, so the proportion is true. **(Test the truth of a proportion)**

23. False

$(7)(12) = (8)(11)$ Cross multiply.

$84 = 88$ The cross products are not equal, so the proportion is false. **(Test the truth of a proportion)**

24. False

$(20)(103) = (11)(187)$ Cross multiply.

$2060 = 2057$ The cross products are not equal, so the proportion is false. **(Test the truth of a proportion)**

25. True

$\left(3\frac{1}{2}\right)(6) = \left(2\frac{1}{3}\right)(9)$ Cross multiply.

$\left(\frac{7}{2}\right)(6) = \left(\frac{7}{3}\right)(9)$ Change mixed numbers to improper fractions.

$21 = 21$ The cross products are equal, so the proportion is true. **(Test the truth of a proportion)**

26. True

$(16.24)(602) = (24.08)(406)$ Cross multiply.

$9,776.48 = 9,776.48$ The cross products are equal, so the proportion is true.
(Test the truth of a proportion)

27. False

$(2.1)(2.3) = (3.2)(1.2)$ Cross multiply.

$4.83 = 3.84$ The cross products are not equal, so the proportion is false.
(Test the truth of a proportion)

28. True

$(0.3)(4) = (1.2)(1)$ Cross multiply.

$1.2 = 1.2$ The cross products are equal, so the proportion is true.
(Test the truth of a proportion)

29. True

$(2)\left(\dfrac{1}{2}\right) = \left(\dfrac{1}{4}\right)(4)$ Cross multiply.

$1 = 1$ The cross products are equal, so the proportion is true.
(Test the truth of a proportion)

30. True

$(5)\left(1\dfrac{1}{15}\right) = \left(2\dfrac{2}{3}\right)(2)$ Cross multiply.

$(5)\left(\dfrac{16}{15}\right) = \left(\dfrac{8}{3}\right)(2)$ Change mixed numbers to improper fractions.

$5\dfrac{1}{3} = 5\dfrac{1}{3}$ The cross products are equal, so the proportion is true.

(Test the truth of a proportion)

31. $(5)\,(x) = (4)\,(3)$

$5x = 12$ Cross multiply.

$x = \dfrac{12}{5}$ Divide both sides of the equation by 5.

$x = 2\dfrac{2}{5}$ Change the improper fraction to a mixed number. **(Solve proportions)**

32. $(25)(x) = (10)(40)$

$25x = 400$ Cross multiply.

$$x = \frac{400}{25}$$ Divide both sides of the equation by 24.

$x = 16$ **(Solve proportions)**

33. $(24)(x) = (20)(18)$

$24x = 360$ Cross multiply.

$$x = \frac{360}{24}$$ Divide both sides of the equation by 24.

$x = 15$ Simplify the improper fraction. **(Solve proportions)**

34. $(15)(x) = (6)(55)$

$15x = 330$

$$x = \frac{330}{15}$$

$x = 22$ **(Solve proportions)**

35. $(10)(x) = \left(\frac{1}{5}\right)(150)$

$10x = 30$ Cross multiply.

$$x = \frac{30}{10}$$ Divide both sides of the equation by 10.

$x = 3$ Simplify the improper fraction. **(Solve proportions)**

36. $\left(\frac{2}{5}\right)(x) = (4)(10)$

$\frac{2}{5}x = 40$ Cross multiply.

$x = (40)\left(\frac{5}{2}\right)$ Multiply both sides of the equation by $\frac{5}{2}$.

$x = 100$ **(Solve proportions)**

37. $(1.25)(x) = (0.5)(5)$

$1.25x = 2.5$ Cross multiply.

$$x = \frac{2.5}{1.25}$$ Divide both sides of the equation by 1.25.

$x = 2$ Divide. **(Solve proportions)**

38. $(0.6)(x) = (12)(5)$

         $0.6x = 60$ Cross multiply.

           $x = 100$ Divide both sides of the equation by 0.6. **(Solve proportions)**

39. $(7.5)(x) = (1.5)(2.5)$

         $7.5x = 3.75$ Cross multiply.

           $x = 0.5$ Divide both sides of the equation by 7.5. **(Solve proportions)**

40. $\left(\dfrac{3}{8}\right)(x) = \left(\dfrac{3}{5}\right)(10)$

         $\dfrac{3}{8}x = 6$ Cross multiply.

           $x = (6)\left(\dfrac{8}{3}\right)$ Multiply both sides of the equation by $\dfrac{8}{3}$.

           $x = 16$ Multiply. **(Solve proportions)**

41. $(2.5)(x) = (8)\left(10\dfrac{1}{2}\right)$ Cross multiply.

        $\left(2\dfrac{1}{2}\right)(x) = \left(\dfrac{1}{3}\right)\left(\dfrac{1}{2}\right)$ Change the decimal to a fraction.

        $\left(\dfrac{5}{2}\right)(x) = \left(\dfrac{1}{6}\right)$ Change the mixed number to an improper fraction.

          $x = \left(\dfrac{1}{6}\right)\left(\dfrac{2}{5}\right)$ Multiply both sides of the equation by $\dfrac{2}{5}$.

          $x = \dfrac{1}{15}$ Multiply. **(Solve proportions)**

42. $\left(3\dfrac{1}{2}\right)(x) = (8)\left(10\dfrac{1}{2}\right)$ Cross multiply.

        $\left(\dfrac{7}{2}\right)(x) = (8)\left(\dfrac{21}{2}\right)$ Change the mixed numbers to improper fractions.

        $\left(\dfrac{7}{2}\right)(x) = 84$ Multiply.

          $x = 24$ Multiply both sides of the equation by $\dfrac{2}{7}$. **(Solve proportions)**

43. $(12)(x) = (0.23)(9)$

$\qquad 12x = 2.07$ Cross multiply.

$\qquad\quad x = 0.1725$ Divide both sides of the equation by 12. **(Solve proportions)**

44. $\dfrac{6 \text{ pages}}{24 \text{ minutes}} = \dfrac{x \text{ pages}}{60 \text{ minutes}}$

$\qquad (24)(x) = (6)(60)$ Cross multiply.

$\qquad\quad 24x = 360$

$\qquad\qquad x = 15 \text{ pages}$ Divide both sides of the equation by 42. **(Word problems using proportions)**

45. $\dfrac{4 \text{ cups of flour}}{3 \text{ tbs. sugar}} = \dfrac{28 \text{ cups flour}}{x \text{ tbs. sugar}}$

$\qquad (4)(x) = (28)(3)$

$\qquad\quad 4x = 84$ Cross multiply.

$\qquad\qquad x = 21 \text{ tbs. of sugar}$ Divide both sides of the equation by 4.
(Word problems using proportions)

46. $\dfrac{15 \text{ engines}}{2 \text{ days}} = \dfrac{x \text{ engines}}{8 \text{ days}}$

$\qquad (2)(x) = (15)(8)$ Cross multiply.

$\qquad\quad 2x = 120$

$\qquad x = 60 \text{ engines}$ Divide both sides of the equation by 2. **(Word problems using proportions)**

47. $\dfrac{\frac{1}{4} \text{ inch}}{3 \text{ miles}} = \dfrac{5 \text{ inches}}{x \text{ miles}}$

$\qquad \left(\dfrac{1}{4}\right)(x) = (3)(5)$ Cross multiply.

$\qquad\quad \dfrac{1}{4}x = 15$

$\qquad\qquad x = 60 \text{ miles}$ **(Word problems using proportions)**

48. $\dfrac{165 \text{ miles}}{3 \text{ hours}} = \dfrac{x \text{ miles}}{8 \text{ hours}}$

$\qquad (3)(x) = (165)(8)$ Cross multiply.

$$3x = 1320$$

$$x = 440 \text{ miles}$$ Divide both sides of the equation by 3. (**Word problems using proportions**)

49. $\dfrac{7 \text{ women}}{5 \text{ men}} = \dfrac{3,500 \text{ women}}{x \text{ men}}$

$$(7)(x) = (3,500)(5)$$ Cross multiply.

$$7x = 17,500$$

$$x = 2,500 \text{ men}$$ Divide both sides of the equation by 7. (**Word problems using proportions**)

50. $\dfrac{\$2,000 \text{ invested}}{\$240 \text{ dividend}} = \dfrac{x \text{ invested}}{\$600 \text{ dividend}}$

$$(240)(x) = (2,000)(600)$$ Cross multiply.

$$240x = 1,200,000$$

$$x = \$5,000 \text{ invested}$$ Divide both sides of the equation by 240. (**Word problems using proportions**)

51. $\dfrac{\frac{1}{4} \text{ inch}}{12° \text{ rise}} = \dfrac{x \text{ inches}}{48° \text{ rise}}$

$$(12)(x) = \left(\dfrac{1}{4}\right)(48)$$ Cross multiply.

$$12x = 12$$

$$x = 1 \text{ inch}$$ Divide both sides of the equation by 12. (**Word problems using proportions**)

52. $\dfrac{3\frac{1}{2} \text{ pounds}}{\$14} = \dfrac{x \text{ pounds}}{\$22}$

$$(14)(x) = \left(3\dfrac{1}{2}\right)(22)$$ Cross multiply.

$$14x = \left(\dfrac{7}{2}\right)(22)$$ Change the mixed number to an improper fraction.

$$x = 5.5 \text{ pounds}$$ Divide both sides of the equation by 14.

$$x = 5\dfrac{1}{2} \text{ pounds}$$ This is the answer in fraction form. (**Word problems using proportions**)

53. $\dfrac{120 \text{ miles}}{3\frac{1}{2} \text{ gallons}} = \dfrac{300 \text{ miles}}{x \text{ gallons}}$

$(120)\,(x) = \left(3\frac{1}{2}\right)(300)$ Cross multiply.

$120x = \left(\dfrac{7}{2}\right)(300)$ Change the mixed number to an improper fraction.

$120x = 1050$

$x = 8.75 \text{ gallons}$ Divide both sides of the equation by 120.

$x = 8\dfrac{3}{4} \text{ gallons}$ This is the answer in fraction form. **(Word problems using proportions)**

54. $\dfrac{x \text{ total population}}{500 \text{ tagged}} = \dfrac{200 \text{ netted}}{4 \text{ tagged}}$

$(4)\,(x) = (500)\,(200)$ Cross multiply.

$4x = 100,000$

$x = 25,000 \text{ fish}$ Divide both sides of the equation by 4. **(Word problems using proportions)**

Grade Yourself

Circle the question numbers that you had incorrect. Then indicate the number of questions you missed. If you answered more than three questions incorrectly, you need to focus on that topic. (If a topic has less than three questions and you had at least one wrong, we suggest you study that topic also. Read your textbook or a review book, or ask your teacher for help.)

Subject: Ratio and Proportion

Topic	Question Numbers	Number Incorrect
Write as a simplified ratio	1, 2, 3, 4, 5, 6, 7, 8, 9, 10, 11	
Write as a simplified rate	12, 13, 14, 15, 16	
Write as a unit rate	17, 18, 19, 20, 21	
Test the truth of a proportion	22, 23, 24, 25, 26, 27, 28, 29, 30	
Solve proportions	31, 32, 33, 34, 35, 36, 37, 38, 39, 40, 41, 42, 43	
Word problems using proportions	44, 45, 46, 47, 48, 49, 50, 51, 52, 53, 54	

Percents

5

Brief Yourself

This chapter reviews the meaning and the usage of percents. It contains questions and answers about converting between fractions, decimals, and percents, solving percent problems both with equations and proportions, and other applications of percents.

Percent means part of 100. As a fraction, a "%" means $\frac{1}{100}$, and as a decimal it means 0.01.

100% is equivalent to 1 or $\frac{100}{100}$. If a percent is less than 100%, it will be equivalent to a fraction or decimal less than 1. If a percent is more than 100%, it will be equivalent to fraction or decimal that is larger than 1.

1%	one percent	1 part out of 100
83%	eighty-three percent	83 parts out of 100
$17\frac{1}{2}\%$	seventeen and $\frac{1}{2}$ percent	$17\frac{1}{2}$ parts out of 100
$\frac{1}{2}\%$	one half percent	$\frac{1}{2}$ of one part out of 100
136%	one hundred thirty-six percent	136 parts out of 100

There are three equivalent ways of writing the same number. One involves the use of fractions, one uses decimals, and one uses percents.

One can say:

"I got $\frac{1}{2}$ of the questions right on my test."

"I got 0.5 of the questions right on my test."

"I got 50% of the questions right on my test."

All of these statements mean the same thing. The formal rules needed to change from one format to another are as follows:

Percent → Decimal

1. % can be replaced by 0.01.

2. Multiply.

$$42\% \to (42)\,(0.01) = 0.42$$

The shortcut for this technique is to drop the % and shift the decimal point two places to the left, adding zeros if necessary.

Remember a percent that is less than 100% is equivalent to a decimal less than 1.

If a percent is in a fraction form, change it to a decimal form before moving the decimal point.

$$\frac{3}{4}\% \to 0.75\% \to (0.75)\,(0.01) \to 0.0075$$

Decimal → Percent

1. Multiply the decimal number by a form of 1. (Remember, since 1 is the multiplicative identity, it will result in an equivalent form.) The form of 1 that should be used is 100%.

$$0.57 \to (0.57)\,(100\%) = 57\%$$

The shortcut is to shift the decimal point two places to the right, add zeros if necessary, and insert the % sign.

Percent → Fraction

1. Replace the % by $\frac{1}{100}$ and multiply.

2. Reduce to lowest terms.

$$46\% \to (46)\left(\frac{1}{100}\right) = \frac{46}{100} = \frac{23}{50}$$

If the percent has a decimal point in it, change it to a fraction before proceeding.

$$1.2 = 1\frac{2}{10}\% = \left(\frac{12}{10}\right)\left(\frac{1}{100}\right) = \frac{3}{250}$$

Fraction → Percent

Multiply the fraction by 100%, change to a mixed number, and reduce if necessary.

Remember: When a fraction is multiplied by a form of 1, the result is equivalent to the original fraction.

Remember: 1 → 100%

Example: $\frac{3}{4}$ becomes $\frac{3}{4} \cdot 100\% = 75\%$

Memorize:

$\dfrac{1}{4}$	0.25	25%
$\dfrac{1}{2}$	0.50	50%
$\dfrac{3}{4}$	0.75	75%
$\dfrac{1}{3}$	0.33...	$33\dfrac{1}{3}\%$
$\dfrac{2}{3}$	0.66...	$66\dfrac{2}{3}\%$

Percent Problems Solved as Equations

To solve a percent problem, translate the problem into an equation and solve.

Key Terms

English	*Mathematics*
is	=
of	$\cdot$ (multiply)
a number	n
what number	n
what percent	P

NOTE: To do arithmetic with percents, the percent must first be changed to either its decimal or fraction equivalent. Some problems will be easier to do with decimals, and some will be easier to do with fractions.

Example:

80% of 45 is what number?

$$(0.80)\ (45)\ =\ n \qquad\qquad 80\% = (80)(0.01) = 0.80$$

$$36\ =\ n$$

70% of what number is 28?

As a decimal

$$(0.70)\ (n)\ =\ 28 \qquad\qquad 0.70\ =\ (70)\left(\frac{1}{100}\right) = \left(\frac{70}{100}\right) = \frac{7}{10}$$

$$\frac{10}{7}\cdot\left(\frac{7}{10}\right)(n)\ =\ 28\cdot\frac{10}{7} \qquad\qquad \text{Solve for } n \text{ by multiplying both sides by } \frac{10}{7}.$$

$$n\ =\ 4\cdot 10$$

$$n\ =\ 40$$

What percent of 60 is 45?

$$(P)\,(60)\;=\;45$$

$$\frac{P\,(60)}{60}\;=\;\frac{45}{60}$$

$$P\;=\;\frac{45}{60}\;=\;\frac{9}{12}\;=\;\frac{3}{4}\qquad\qquad\left(\frac{3}{4}\right)(100\%)\;=\;75\%$$

$$=75\%$$

NOTE: When looking for percent, use P as the variable. It will serve as a reminder to change the decimal or fraction answer into a percent.

To solve percent problems by proportion, use the following formula:

Percent % of Base is Amount

$$\qquad P \qquad\qquad B \qquad\quad A$$

$$\frac{A}{B}\;=\;\frac{P}{100}$$

Cross multiply to solve for the unknown.

80% of 45 is what number?

$$\frac{x}{45}\;=\;\frac{80}{100}$$

$$(100)\,(x)\;=\;(45)\,(80)$$

$$(100)\,(x)\;=\;3600$$

$$\frac{100x}{100}\;=\;\frac{3600}{100}$$

$$x\;=\;36$$

70% of what number is 28?

$$\frac{28}{x}\;=\;\frac{70}{100}$$

$$(70)\,(x)\;=\;(28)\,(100)$$

$$70x\;=\;2800$$

$$\frac{70x}{70}\;=\;\frac{2800}{70}$$

$$x\;=\;40$$

What percent of 60 is 45?

$$\frac{45}{60} = \frac{P}{100}$$

$$60P = (45)(100)$$

$$60P = 4500$$

$$P = 75\%$$

To solve application problems, it is easier if the problem is translated into the form _____% of _____is _____. After it is in this form, check to make sure that the rate (%) and the amount are referring to the same quantity.

Example:

If John gets 80% of the questions correct on his math test, how many questions were on the test if he got 15 correct?

80% of _____ is 15.

NOTE: 80% represents the rate at which he gets the questions *correct*.

15 is the number of questions he gets *correct*.

These measure the same thing but in different ways.

If John gets 80% of the questions correct on his math test, how many questions were on the test if he got 5 wrong?

20% of _____ is 5.

NOTE: 80% represents the rate at which he gets the questions *correct*.

5 represents the number of questions he got *wrong*.

The two numbers used in the problem must measure the same thing. Since there is no way to know the number of correct answers, the percent of wrong answers must be used. If 80% of the answers are correct, then 20% of them must be wrong.

There are many business applications of percent problems:

Sales tax is collected by the government. It is added on to the cost of an item.

Commission is the amount of money a person receives. It is often based on a percent of the total goods sold in a set time period.

Discount is the difference between the original price of an item and its sale price.

Markup is the amount the store adds to the cost of an item.

Percent increase/decrease is the change based on the original amount.

Simple interest is the amount of money owed or paid for borrowing or saving money.

Use the formula: $I = P \cdot R \cdot T$ where

I = Interest

P = Principal and borrowed amount saved.

R = Rate (%)

T = Time (in years)

 # Test Yourself

In problems 1 – 25, find the equivalent forms.

	Fraction	Decimal	Percent
1.	$\frac{3}{8}$		
2.		0.52	
3.			90%
4.	$\frac{4}{5}$		
5.		0.6	
6.			15%
7.	$7\frac{1}{4}$		
8.		0.035	
9.			4.5%
10.	$\frac{1}{16}$		
11.		1.675	
12.			0.6%
13.	$\frac{5}{4}$		

	Fraction	Decimal	Percent
14.		0.008	
15.			$\frac{3}{5}\%$
16.	9		
17.		0.745	
18.			$18\frac{1}{3}\%$
19.	$\frac{13}{100}$		
20.		1.23	
21.			$24\frac{7}{8}\%$
22.	$\frac{2}{9}$		
23.		12	
24.			0.05%
25.	$1\frac{7}{8}$		

Solve problems 26 – 46, using either the equation method or the proportion method.

26. 32% of what number is 95?

27. 1.3 is what percent of 6.5?

28. 78% of 36 is what number?

29. 18 is what percent of 54?

30. $1\frac{2}{3}$ is what percent of $8\frac{1}{3}$?

31. 48% of 40 is what number?

32. $1\frac{1}{2}$ % of 2,500 is what number?

33. $16\frac{2}{3}$ % of 327 is what number?

34. What percent of $2\frac{2}{3}$ is $\frac{8}{15}$?

35. 3.2% of 0.7 is what number?

36. What percent of 7.5 is 3?

37. $130\frac{1}{2}$ % of 245 is what number?

38. 56 is what percent of 48?

39. What number is 0.5% of 0.5?

40. $4\frac{1}{8}$ is what percent of $6\frac{7}{8}$?

41. 43% of what number is 43?

42. $5\frac{4}{7}$ % of $\frac{35}{39}$ is what number?

43. 57.1% of what number is 1,142?

44. $62\frac{1}{2}$ % of what number is 200?

45. 15.4% of what number is 130.9?

46. $\frac{3}{4}$ % of 219 is what number?

47. Penny spends $450 each month on food for her family. If her monthly income is $1,800, what

percent of her income is spent on food per month?

48. If Jeanette answered 37 items correctly on a test and received a score of 74%, how many items were on the test?

49. On a 20-item practice test, how many questions must Dana answer to receive a score of 80%?

50. Don expects 1.04% of the tires manufactured at his factory to be defective. If a run produced 27,500 tires, how many can Don expect to be defective?

51. How much can Melanie expect to pay for a $32.00 shirt if it is discounted by 15%?

52. Sabine's weekly salary is $320 plus 5% commission on her sales. If Sabine's sales for the week totaled $800, how much did she earn in total for the week?

53. In a recent tennis match, Kevin attempted 25 service aces. If 17 of them were good, what percent of his attempts were successful?

54. Four of Mrs. Marshall's mathematics students scored 100% on the exam. If this represents 16% of the class, how many students are in Mrs. Marshall's class?

55. During a sale, Johanna found a blouse that originally cost $20 marked down to $12. What was the percent of decrease for the price of the blouse?

56. After taking a special reading course, Eleanor was able to increase her reading rate from 250 words per minute to 360 words per minute. What was the percent increase in her reading rate?

57. Ron bought four pairs of socks at $2.49 for each pair. If the sales tax is 6%, how much did he need to pay for the socks, including the tax?

58. Betsy earns 1.5% commission for selling real estate. If her commission for this month was $1,710 and she sold only one house, what was the cost of the house she sold?

59. By changing her driving habits, Jean was able to increase her car's rate of mileage per gallon from 19.5 to 23.7. What is the percent increase?

60. Of the 10,000 freshman students, 2,000 register for a psychology class. What percent of the freshman register for psychology?

61. Sue paid $979.60 in sales tax on a new car. If the cost of the car was $15,800, what was the rate of sales tax?

62. The sales tax on a purchase is $66. This represents a sales tax rate of 5.5%. What was the price of the purchase?

63. Gloria saves 6% of her salary. If she saves $900 one year, what is her annual salary?

64. On a 50-question math test, Carol answered 78% of the questions correctly. How many questions did she answer incorrectly?

65. Eva bought an alarm system from a wholesaler for $1,500. If before she sells it in her store she marks up the price by 8.2%, what is the new selling price of the alarm system?

66. If Sue invests $1,500 in an account that earns 8.5% simple interest, how much will she have at the end of four years?

67. How much interest will Gerry have to repay on a six-month loan of $30,000 at 8.6% simple interest?

68. Jeanne sold $7,826 of jewelry this week. How much did she make if her commission rate is 5.5%?

69. Katie bought a new car for $15,460. She paid 20% as a down payment. If the balance is to be paid in 48 equal monthly installments, how much must she pay each month (round to nearest cent)?

70. Barbara earns 12% commission on each lab manual she sells. If she sells 1,200 manuals at $9.95 each, find her commission.

Check Yourself

NOTE: Problems 1 – 25 can be done in a variety of ways. Any of the rules given for changing fractions, decimals, and percents can be used.

1. decimal   0.375   Divide $8\overline{)3.000}$.

   percent   37.5%   Shift decimal point 2 places right.
   (Percent, fraction, decimal equivalents)

2. fraction $\dfrac{13}{25}$ Reduce $\dfrac{52}{100}$.

   percent   52%   Shift decimal point 2 places right.
   (Percent, fraction, decimal equivalents)

3. fraction $\dfrac{9}{10}$ Reduce $\dfrac{90}{100}$.

   decimal   0.9   Divide $10\overline{)9.0}$.
   (Percent, fraction, decimal equivalents)

4. decimal   0.8   Divide $5\overline{)4.0}$.

   percent   80%   Shift decimal point 2 places right.
   (Percent, fraction, decimal equivalents)

5. fraction $\dfrac{3}{5}$ Reduce $\dfrac{6}{10}$.

   percent   60%   Shift decimal point 2 places right.
   (Percent, fraction, decimal equivalents)

6. fraction $\dfrac{3}{20}$ Reduce $\dfrac{15}{100}$.

     decimal   0.15   Shift decimal point 2 places left.
(Percent, fraction, decimal equivalents)

7. decimal   7.25   Write 1/4 as 0.25.
     percent   725%   Shift decimal point 2 places right.
(Percent, fraction, decimal equivalents)

8. fraction $\dfrac{7}{200}$ Reduce $\dfrac{35}{1000}$.

     percent   3.5%   Shift decimal point 2 places right.
(Percent, fraction, decimal equivalents)

9. decimal   0.045   Shift decimal point 2 places left.

     fraction $\dfrac{9}{200}$ Reduce $\dfrac{45}{1000}$.

(Percent, fraction, decimal equivalents)

10. decimal   0.0625   Divide $16\overline{)1.0000}$.

     percent   6.25%   Shift decimal point 2 places right.
(Percent, fraction, decimal equivalents)

11. percent   167.5%   Shift decimal point 2 places right.

     fraction $1\dfrac{27}{40}$ Reduce $\dfrac{675}{1000}$.

(Percent, fraction, decimal equivalents)

12. decimal   0.006   Shift decimal point 2 places left.

     fraction $\dfrac{3}{500}$ Reduce $\dfrac{6}{1000}$.

(Percent, fraction, decimal equivalents)

13. decimal   1.25   Divide $4\overline{)5.00}$.

     percent   125%   Shift decimal point 2 places right.
(Percent, fraction, decimal equivalents)

14. fraction $\dfrac{1}{125}$ Reduce $\dfrac{8}{1000}$.

     percent   0.8%   Shift decimal point 2 places right.
(Percent, fraction, decimal equivalents)

15. decimal   0.006 $\dfrac{3}{5}\%$ can be rewritten 0.6%, then shift decimal point 2 places left.

     fraction $\dfrac{3}{500}$ Reduce $\dfrac{6}{1000}$.

(Percent, fraction, decimal equivalents)

16. decimal   9   Whole number.
     percent   900%   Shift decimal point 2 places right.
(Percent, fraction, decimal equivalents)

17. percent   74.5%   Shift decimal point 2 places right.

     fraction $\dfrac{149}{200}$ Reduce $\dfrac{745}{1000}$.

                                (Percent, fraction, decimal equivalents)

18. fraction $\dfrac{11}{60}$ Change to improper $\dfrac{55}{3}$ and multiply by $\dfrac{1}{100}$.

     decimal   0.183. . .   Write $18\dfrac{1}{3}$ as 18.333. . . and shift decimal point 2 places left.

                                (Percent, fraction, decimal equivalents)

19. decimal   0.13   Divide $100\overline{)13.00}$.
     percent   13%   Shift decimal point 2 places right.
                                (Percent, fraction, decimal equivalents)

20. percent   123%   Shift decimal point 2 places right.

     fraction $1\dfrac{23}{100}$

                                (Percent, fraction, decimal equivalents)

21. decimal   0.24875   Divide $8\overline{)7.000}$, then shift decimal point 2 places left.

     fraction $\dfrac{199}{800}$ Reduce $\dfrac{24875}{100000}$.

                                (Percent, fraction, decimal equivalents)

22. decimal $0.22\dfrac{2}{9}$ Divide $9\overline{)2.00}$.

     percent $22\dfrac{2}{9}\%$ Shift decimal point 2 places right.

                                (Percent, fraction, decimal equivalents)

23. fraction   12   Whole number.
     percent   1200%   Shift decimal point 2 places right.
                                (Percent, fraction, decimal equivalents)

24. decimal   0.0005   Shift decimal point 2 places left.

     fraction $\dfrac{1}{2000}$ Reduce $\dfrac{5}{10000}$.

                                (Percent, fraction, decimal equivalents)

25. decimal   1.875   Divide $8\overline{)7.000}$.

     percent   187.5%   Shift decimal point 2 places right.
                                (Percent, fraction, decimal equivalents)

Problems 26 – 46 can generally be solved two ways: as equations (changing the percent to a fraction or changing the percent to a decimal) or as proportions.

26. $(0.32)\,n = 95$ Solving as an equation, change the percent to a decimal by shifting the decimal point 2 places left.

$$\dfrac{0.32n}{0.32} = \dfrac{95}{0.32}$$

$n = 296.875$ Solve for n by dividing both sides by 0.32. **(Solving percent problems)**

27. $1.3 = (P)(6.5)$ Translate into an equation.

$\dfrac{1.3}{6.5} = \dfrac{(P)(6.5)}{(6.5)}$ Solve for P by dividing both sides by 6.5.

$0.2 = P$ Change the decimal into a percent by shifting the decimal point 2 places to the right.

$20\% = P$ **(Solving percent problems)**

28. $\dfrac{n}{36} = \dfrac{78}{100}$ Set up as a proportion.

$100n = 2808$ Cross multiply.

$\dfrac{100n}{100} = \dfrac{2808}{100}$ Solve for n by dividing both sides by 100.

$n = 28\dfrac{2}{25}$ Express answer as either a fraction or a decimal.

$n = 28.08$ **(Solving percent problems)**

29. $\dfrac{18}{54} = \dfrac{P}{100}$ When solving for a percent using the proportion method, remember when P is isolated it will already be in percent form.

$54P = 1800$

$\dfrac{54P}{54} = \dfrac{1800}{54}$

$P = 33\dfrac{1}{3}\%$ **(Solving percent problems)**

30. $1\dfrac{2}{3} = (P)\left(8\dfrac{1}{3}\right)$ Because the fractions are equivalent to repeating decimals, fractions should be used.

$\dfrac{5}{3} = (P)\left(\dfrac{25}{3}\right)$

$\dfrac{3}{25} \cdot \dfrac{5}{3} = (P)\left(\dfrac{25}{3}\right)\left(\dfrac{3}{25}\right)$ Solve for P by multiplying by the reciprocal.

$\dfrac{1}{5} = P$ Change the fraction into a percent.

$20\% = P$ **(Solving percent problems)**

31. $(0.48)(40) = n$ Change the percent to a decimal.
Translate into an equation.

$19.2 = n$ Solve for n by multiplying. **(Solving percent problems)**

32. $1\frac{1}{2}\%$ can be written as 1.5%   Change the percent to a decimal.

$(0.015)\,(2500)\ =\ n$

$37.5\ =\ n$ **(Solving percent problems)**

33. $16\frac{2}{3}\%$ becomes $\frac{50}{3}\cdot\frac{1}{100}=\frac{1}{6}$ Change to improper fraction, $\frac{50}{3}$, then multiply by $\frac{1}{100}$.

$\left(\frac{1}{6}\right)(327)\ =\ n$ Translate into an equation.

$54\frac{1}{2}\ =\ n$ **(Solving percent problems)**

34. $(P)\left(2\frac{2}{3}\right)\ =\ \frac{8}{15}$ Translate into an equation.

$(P)\frac{8}{3}\ =\ \frac{8}{15}$

$\frac{3}{8}\cdot\frac{8}{3}\cdot P\ =\ \frac{8}{15}\cdot\frac{3}{8}$ Solve for P by multiplying both sides by $\frac{3}{8}$.

$P\ =\ \frac{1}{5}$ Change the fraction into a percent.

$P\ =\ 20\,\%$ **(Solving percent problems)**

35. $(0.032)\,(0.7)\ =\ n$ Change the percent to a decimal.

$0.0224\ =\ n$ Solve for n by multiplying. **(Solving percent problems)**

36. $\frac{3}{7.5}\ =\ \frac{P}{100}$ Set up proportion.

$7.5P\ =\ (3)\,(100)$ Cross multiply.

$7.5P\ =\ 300$ Solve for P by dividing by 7.5.

$P\ =\ 40\%$ **(Solving percent problems)**

37. $\dfrac{n}{245}\ =\ \dfrac{130\frac{1}{2}}{100}$ Set up proportion.

$100n\ =\ \left(130\frac{1}{2}\right)(245)$ Cross multiply.

$100n\ =\ \dfrac{(261)}{2}\cdot\dfrac{(245)}{1}$ Change mixed number to an improper fraction.

$$100n = \frac{63945}{2}$$

$$\left(\frac{1}{100}\right)(100)\,n = \frac{63945}{2} \cdot \frac{1}{100}$$

$$n = 319\frac{29}{40} \quad \textbf{(Solving percent problems)}$$

38. $\dfrac{56}{48} = \dfrac{P}{100}$ Set up proportion.

$48P = 5600$ Cross multiply.

$P = 116\dfrac{2}{3}\%$ Divide both sides by 48. **(Solving percent problems)**

39. $n = (0.005)(0.5)$ Change percent to a decimal.
Translate to an equation.

$n = 0.0025$ Solve for n. **(Solving percent problems)**

40. $\dfrac{4\frac{1}{8}}{6\frac{7}{8}} = \dfrac{P}{100}$ Set up proportion.

$\left(4\frac{1}{8}\right)(100) = \left(6\frac{7}{8}\right)(P)$ Cross multiply.

$\left(\frac{33}{8}\right)(100) = \left(\frac{55}{8}\right)(P)$ Change mixed numbers to improper fractions.

$\dfrac{8}{55} \cdot \dfrac{33}{8} \cdot \dfrac{100}{1} = \dfrac{8}{55} \cdot \dfrac{55}{8} \cdot P$ Multiply both sides of the equation by $\dfrac{8}{55}$ to isolate P.

$60\% = P$ **(Solving percent problems)**

41. $\dfrac{43}{100} \cdot n = 43$ Change percent into a fraction.

$\dfrac{100}{43} \cdot \dfrac{43}{100} \cdot n = 43 \cdot \dfrac{100}{43}$ Multiply both sides of the equation by $\dfrac{100}{43}$.

$n = 100$ **(Solving percent problems)**

42. $\dfrac{n}{\frac{35}{39}} = \dfrac{5\frac{4}{7}}{100}$ Set up proportion.

$100n = \left(\dfrac{35}{39}\right)\left(\dfrac{39}{7}\right)$ Cross multiply.

$$100n = 5$$

$$n = \frac{1}{20}$$ Multiply both sides by $\frac{1}{100}$ to solve for n. (**Solving percent problems**)

43. $0.571n = 1142$ Change percent to decimal.

$$\frac{0.571n}{0.571} = \frac{1142}{0.571}$$ Divide both sides by 0.571 to solve for n.

$$n = 2000$$ (**Solving percent problems**)

44. $\frac{5}{8}n = 200$ Change the percent to a fraction.

$$\frac{8}{5} \cdot \frac{5}{8}n = 200 \cdot \frac{8}{5}$$ Multiply both sides by $\frac{8}{5}$ to solve for n.

$$n = 320$$ (**Solving percent problems**)

45. $0.154n = 130.9$ Change the percent to a decimal.

$$\frac{0.154n}{0.154} = \frac{130.9}{0.154}$$ Divide both sides by 0.154 to solve for n.

$$n = 850$$ (**Solving percent problems**)

46. $\frac{n}{219} = \frac{\frac{3}{4}}{100}$ Set up proportion.

$$100n = \left(\frac{3}{4}\right)(219)$$ Cross multiply.

$$100n = \frac{657}{4}$$

$$\frac{1}{100} \cdot 100 \cdot n = \frac{657}{4} \cdot \frac{1}{100}$$

$$n = 1\frac{257}{400}$$ (**Solving percent problems**)

47.   What % of 1,800 is 450?   Rewrite the problem as an equation.

$$(1800)(P) = 450$$

$$\frac{1800p}{1800} = \frac{450}{1800}$$ Divide both sides by 1800 to solve for P.

$$P = 0.25$$

$$P = 25\%$$ Change the decimal to a percent. (**Applications of percents**)

48.   74% of what number is 37?

$$0.74n = 37$$

$$\frac{0.74n}{0.74} = \frac{37}{0.74}$$

$n = 50$ questions **(Applications of percents)**

49.   80% of 20 is what number?

$$(0.80)\,(20) = n$$

16 questions $= n$ **(Applications of percents)**

50.   1.04% of 27,500 is what number?

$$(0.0104)\,(27,500) = n$$

$$286 = n$$

$n = 286$ defective tires **(Applications of percents)**

51.   15% of 32 is what number?

$$(0.15)\,(32) = n$$

$\$4.80 = n$ amount saved

$32.00 - 4.80 = \$27.20$ sale price of shirt **(Applications of percents)**

52.   5% of 800 is what number?

$$(0.05)(800) = n$$

$\$40 = n$ earned in commission

$320 + 40 = \$360$ total earnings for a week **(Applications of percents)**

53.   What percent of 25 is 17?

$$(P)\,(25) = 17$$

$$\frac{P\,(25)}{25} = \frac{17}{25}$$

$$P = 0.68$$

$$P = 68\% \quad \textbf{(Applications of percents)}$$

54.   16% of what number is 4?

$$(0.16)\,n = 4$$

$$\frac{(0.16)\,n}{0.16} = \frac{4}{0.16}$$

$$n = 25 \text{ students } \textbf{(Applications of percents)}$$

55. What % of 20 is 8?

$$(P)\,(20) = 8$$

$$\frac{(P)\,(20)}{20} = \frac{8}{20}$$

$$P = 0.4$$

$$P = 40\% \text{ decrease } \textbf{(Applications of percents)}$$

56. What % of 250 is 110?

$$(P)\,(250) = 110$$

$$\frac{(P)\,(250)}{250} = \frac{110}{250}$$

$$P = 0.44$$

$$P = 44\% \text{ increase } \textbf{(Applications of percents)}$$

57. $(2.49)\,(4) = 9.96$ cost of socks

6% of 9.96 is what number?

$$(0.06)\,(9.96) = n$$

$$0.5976 = n$$

$$0.60 = n \text{ rounded to the nearest penny}$$

9.96 + .60 socks plus tax

10.56 total cost **(Applications of percents)**

58. 1.5% of what number is 1,710?

$$(0.015)\,n = 1710$$

$$\frac{0.015n}{0.015} = \frac{1710}{0.015}$$

$$n = \$114{,}000 \text{ cost of the house } \textbf{(Applications of percents)}$$

59. What % of 19.5 is 4.2?

$$(P)\,(19.5) = 4.2$$

$$\frac{(P)\,(19.5)}{19.5} = \frac{4.2}{19.5}$$

$$P = 21\frac{7}{13}\,\% \quad \textbf{(Applications of percents)}$$

60. What % of 10,000 is 2,000?

$$(P)\,(10,000) = 2,000$$

$$\frac{(P)\,(10,000)}{10,000} = \frac{2,000}{10,000}$$

$$P = \frac{1}{5}$$

$$P = 0.2$$

$$P = 20\,\%\ \text{of the freshman class takes psychology}\ \textbf{(Applications of percents)}$$

61. What % of $15,800 is $979.60?

$$(P)\,(15,800) = 979.60$$

$$\frac{(P)\,(15,800)}{15,800} = \frac{979.60}{15,800}$$

$$P = 0.062$$

$$P = 6.2\,\%\ \text{sales tax}\ \textbf{(Applications of percents)}$$

62. 5.5% of what number is $66?

$$(0.055)\,n = 66$$

$$\frac{0.055n}{0.055} = \frac{66}{0.055}$$

$$n = \$1200 \quad \textbf{(Applications of percents)}$$

63. 6% of what number is 900?

$$(0.06)\,n = 900$$

$$\frac{(0.06)\,n}{0.06} = \frac{900}{0.06}$$

$$n = \$15,000\ \text{annual salary}\ \textbf{(Applications of percents)}$$

64. 78% of 50 is what number?

$$(0.78)\,(50) = n$$

$$n = 39\ \textbf{(Applications of percents)}$$

65. 8.2% of $1500 is what number?

    $(0.082)(1500) = n$

    $\qquad \$123 = n$ amount to be added to cost

    $\$1500 + \$123 = \$1623$ selling price **(Applications of percents)**

66. $I = p \cdot r \cdot t$

    $I = (1500)(0.085)(4)$

    $I = 510$

    Total investment $= \$1500 + \$510$

    $2,010 **(Applications of percents)**

67. $I = (30,000)(0.086)(0.5)$ $\quad$ 6 months $= \dfrac{1}{2}$ year $= 0.5$ year

    $I = \$1,290$ interest owed **(Applications of percents)**

68. 5.5% of $7,826 is what number?

    $(0.055)(7,826) = n$

    $\qquad \$430.43 = n$ amount earned **(Applications of percents)**

69. 20% of 15,460 is what number?

    $(0.20)(15,460) = n$

    $\qquad \$3,092 = n$ down payment

    $\$15,460 - \$3,092 = \$12,368$ still owed

    $\$12,368 \div 48 = \$257.67$ must be paid each month. **(Applications of percents)**

70. $(1200)(9.95) = \$11,940$ total collected from sale

    12% of 11,940 is what number?

    $(0.12)(11,940) = n$

    $1,432.80 amount earned **(Applications of percents)**

Grade Yourself

Circle the question numbers that you had incorrect. Then indicate the number of questions you missed. If you answered more than three questions incorrectly, you need to focus on that topic. (If a topic has less than three questions and you had at least one wrong, we suggest you study that topic also. Read your textbook or a review book, or ask your teacher for help.)

Subject: Percents

Topic	Question Numbers	Number Incorrect
Percent, fraction, decimal equivalents	1, 2, 3, 4, 5, 6, 7, 8, 9, 10, 11, 12, 13, 14, 15, 16, 17, 18, 19, 20, 21, 22, 23, 24, 25	
Solving percent problems	26, 27, 28, 29, 30, 31, 32, 33, 34, 35, 36, 37, 38, 39, 40, 41, 42, 43, 44, 45, 46	
Applications of percents	47, 48, 49, 50, 51, 52, 53, 54, 55, 56, 57, 58, 59, 60, 61, 62, 63, 64, 65, 66, 67, 68, 69, 70	

Measurement and Geometry

6

Brief Yourself

This chapter reviews conversions within both the English and the metric systems of measurement. It will also review the basic geometric concepts of perimeter, area, surface area, and volume.

English System

The following conversions should be memorized:

Length	12 inches (in.)	=	1 foot (ft)
	3 feet	=	1 yard (yd)
	5280 feet	=	1 mile (mi)
Weight	16 ounces (oz)	=	1 pound (lb)
	2000 pounds	=	1 ton
Volume	2 cups (c)	=	1 pint (pt)
	2 pints	=	1 quart (qt)
	4 quarts	=	1 gallon (gal)
Time	60 seconds (sec)	=	1 minute (min)
	60 minutes	=	1 hour (hr)
	24 hours	=	1 day
	7 days	=	1 week

Each of these conversions, when put in a fraction form, is equivalent to 1. For example, $\frac{12 \text{ inches}}{1 \text{ foot}} = 1$

To convert from one unit to another:

— Multiply the measurement by a form of 1. Put the unit that you are converting to in the numerator and the unit you want to eliminate in the denominator.
— Multiply using the appropriate rules.

Example: 124 pt = _____ qt

$$124 \text{ pt} \cdot \frac{1 \text{ qt}}{2 \text{ pt}} = 62 \text{ qt}$$

Sometimes it is necessary to multiply by more than one form of 1 to get the appropriate unit in the answer.

Example: 1200 sec = _____ day

$$1200 \text{ sec} \cdot \frac{1 \text{ min}}{60 \text{ sec}} \cdot \frac{1 \text{ hr}}{60 \text{ min}} \cdot \frac{1 \text{ day}}{24 \text{ hr}} = \frac{1200 \text{ day}}{(60)(60)(24)} = \frac{1200}{86,400} = \frac{12}{864} = \frac{1}{72} \text{ day}$$

Metric System

Basic Unit

Length	meter	m
Weight	gram	g
Volume	liter	L

All others units are named in terms of these basic units along with a prefix.

Kilo	hecto	deka	basic unit	deci	centi	milli
k	h	dk		d	c	m
1000	100	10	1	$\frac{1}{10}$	$\frac{1}{100}$	$\frac{1}{1000}$

Because the metric system is built on the base 10 place value system, the conversions from one unit to another can be made by shifting the decimal point.

When converting from one unit to another, shift the decimal point the same number of places and in the same direction as the number of places and direction it takes you to go from the original prefix to the new one. Add zeros if necessary.

Example:

23 hg = _____ dg   It takes three shifts to the right to get from hectogram to decigram. Therefore,
23.hg = 23,000 dg   move the decimal point three places to the right.

NOTE: Metrics can also be converted by multiplying by a form of 1, as in the English system. However, it is quicker to do it by shifting decimal points.

23 hg = ___ dg

$$23 \text{ hg} \cdot \frac{100 \text{ g}}{1 \text{ hg}} \cdot \frac{10 \text{ dg}}{1 \text{ g}}$$

(23)(100)(10) dg

23,000 dg

English-to-Metric Conversions

English		Metric		Metric		English
1 in.	=	2.54 cm		1 m	=	39.4 in.
1 mi	=	1.61 km		1 km	=	0.6 mi
1 lb	=	454 g		1 kg	=	2.2 lb
1 qt	=	946 ml		1 L	=	1.06 qt

— Multiply the measurement by a form of 1. Put the unit that you are converting to in the numerator and the unit you want to eliminate in the denominator.

— Multiply using the appropriate rules.

Example: 30 in = ___ cm

$$30 \text{ in.} \cdot \frac{2.54 \text{ cm}}{1 \text{ in.}} = (30)(2.54) \text{ cm} = 76.2 \text{ cm}$$

Operations with Measurements

Addition — Write the measurements vertically with like units aligned.
— Add like units together.
— Simplify the answer if possible, using the appropriate conversion.

Example:

```
   3 gal   2 qt
+  5 gal   3 qt
   8 gal   5 qt    5 qt = 4 qt + 1 qt   (4 qt = 1 gal)
                         1 gal + 1 qt

   9 gal   1 qt
```

Subtraction — Same rules as addition, except subtract like units.
— Use the appropriate conversion if borrowing is necessary.

Example:

```
   8 ft 3 in.  =   7 ft + 1 ft + 3 in.
-  2 ft 5 in.  = - 2 ft           5 in.


                   7 ft + 12 in. + 3 in.
                 - 2 ft            5 in.


                   7 ft   15 in.
              ±    2 ft    5 in.
                   5 ft   10 in.
```

Multiplication — Multiply each part of the measurement by the multiplier.
— Simplify the answers if possible by using the appropriate conversions.

Example:

```
   2 hr 6 min
×       11
  22 hr 66 min = 22 hr 60 min + 6 min = 23 hr 6 min
```

Division — Divide each part of the measurement separately, beginning at the left.
— Convert any remainder to the next smaller unit.
— Continue to divide until all parts have been divided.

Example:

$$
\begin{array}{r}
2\ \text{yd}\quad\ 1\ \text{ft}\qquad 7\ \text{in.} \\
7\overline{)17\ \text{yd}\quad 2\ \text{ft}\qquad 1\ \text{in.}} \\
-14\phantom{\ \text{yd}\quad 2\ \text{ft}\qquad 1\ \text{in.}}
\end{array}
$$

3 yd = $\underline{9\ \text{ft}}$

11 ft

$\underline{7}$

4 ft = $\underline{48\ \text{in.}}$

49 in.

$\underline{49\ \text{in.}}$

Perimeter is the distance around a polygon. This distance is found by finding the sum of the lengths of all the sides. Perimeter is expressed in linear units. (Inches, feet, meters, etc.)

NOTE: Before finding this sum, all sides must be converted into the same unit, if necessary.

Circumference is the distance around a circle. Circumference is also expressed in linear units.

Area is the amount of surface inside a two dimensional region. All lengths must be expressed in the same unit in order to find the area. Area is expressed in square units. (sq ft, sq m, sq cm, etc.)

Volume is the amount of space contained inside a three-dimensional region. All lengths must be expressed in the same unit in order to find the volume. Volume is expressed in cubic units. (cu in., cu ft, cu cm, etc.)

Geometric Shapes

NOTE: In the formulas stated below, *P* stands for perimeter, *A* for area, and *V* for volume.

Square

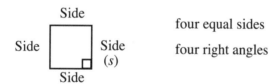

four equal sides

four right angles

$P = 4s$

$A = s^2$

Rectangle

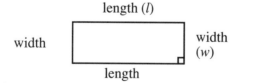

opposite sides are parallel and of equal length

four right angles

$P = l + w + l + w$

$A = l \cdot w$

Parallelogram

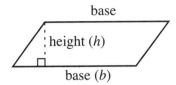

opposite sides are parallel and of equal length

P = sum of four sides

$A = b \cdot h$

Trapezoid

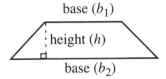

one pair of parallel sides called bases

P = sum of four sides

$A = \frac{1}{2} h (b_1 + b_2)$

Triangle

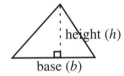

three sides

P = sum of sides

$A = \frac{1}{2} \cdot b \cdot h$

Circle

$C = 2\pi r$ Note: $\pi \approx \frac{22}{7}$ or 3.14

$A = \pi r^2$

Cube

each face is a square
all edge lengths are equal

$V = e^3$

Rectangular Solid

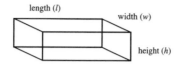

top and bottom are the same size rectangles
left and right are the same size rectangles
front and back are the same size rectangles

$$V = l \cdot w \cdot h$$

Cylinder

top and bottom are circles with radius (r)

$$V = \pi r^2 h$$

Cone

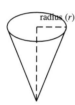

top is a circle with radius (r)

$$V = \frac{1}{3} \pi r^2 h$$

Rectangular Pyramid

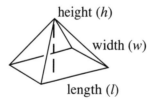

bottom is a rectangle
the four sides are triangles

$$V = \frac{1}{3} (l \cdot w \cdot h)$$

Sphere

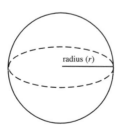

$$V = \frac{4}{3} \pi r^3$$

Test Yourself

In problems 1 – 20, convert each amount to the indicated unit.

1. 12 yd = _____ft

2. 3 lb = _____oz

3. 6 pt = _____cup

4. 25 ft = _____yd

5. 9 pt = _____qt

6. $2\frac{2}{3}$ yd = _____ ft

7. 40 sec = _____ min

8. 90 oz = _____lb

9. 7 wk = _____days

10. 3.5 qt = _____cup

11. $3\frac{1}{2}$ hr = _____min

12. 20 cup = _____gal

13. $\frac{3}{4}$ lb = _____oz

14. 5.9 qt = _____pt

15. 150 in. = _____ft

16. 4.9 ton = _____lb

17. 10 hr = _____day

18. 176 oz = _____lb

19. $2\frac{3}{4}$ gal = _____pt

20. 3.8 mi = _____ft

In problems 21 – 36, convert each amount to the indicated unit.

21. 600 mg = _____g

22. 675 cm = _____km

23. 0.5 hm = _____cm

24. 609 g = _____dg

25. 0.6 mg = _____g

26. 5 mm = _____m

27. 6.08 mL = _____L

28. 9.45 dkL = _____cL

29. 4 km = _____m

30. 4000 mg = _____hg

31. $1\frac{5}{8}$ cm = _____mm

32. 6.019 km = _____m

33. 0.73 kg = _____g

34. 1080mg = _____dg

35. 4200 dL = _____mL

36. 14.92 kL = _____L

In problems 37 – 51, convert each amount to the indicated unit.

37. 2L = _____qt

38. 12 oz = _____g

39. 15kg = _____lb

40. 1 lb = _____g

41. 2km = _____ft

42. 600m = _____yd

43. 6.3L = _____qt

44. 20 yd = _____m

45. 75 ft = _____m

46. 8 lb = _____g

47. 0.85 cup = _____mL

48. 12L = _____qt

49. 27.5 lb = _____ kg

50. 20 gal = _____ L

51. 15 mi = _____ km

In problems 52 – 59, perform the indicated operation. Be sure to simplify each answer.

52. (6ft 8 in.) · 2

53. (5 lb 7 oz.) – (3 lb 10 oz)

54. (12 yd 11 in.) + (15 yd 2 ft 7 in.)

55. (17 min 4 sec) ÷ 8

56. 8 yd - (5ft 3 in.)

57. (2 gal 3 qt) + (8 gal 3 qt)

58. (17 lb 4 oz) ÷ 6

59. (8 gal 3 qt) · 4

60. A trip takes 11 hours 15 minutes by car. If there are five people who want to share the driving equally, how long should each person drive?

61. To fence his back yard, Tony estimates it will take 32 yards of fencing material. Will two rolls of fencing be enough if each roll contains a 50 foot length of material?

62. To frame a door, Anne needs three pieces of molding. Two pieces measure 6 feet 8 inches, and the third piece measures 3 feet 2 inches. If these pieces are cut from a board that measures 17 feet 2 inches, how much will she have left?

63. One can of green beans weighs $8\frac{1}{2}$ ounces. How much do eight cans weigh?

64. Mark has a part-time job after school. He works 3 hours 15 minutes each afternoon for five days a week and then 7 hours 30 minutes on Saturday. How many hours does Mark work each week?

65. If Pat wants to cut an 8 foot 9 inch pipe into five equal pieces, how long should each piece be cut?

66. On Lizzy's seventh birthday, she was 3 feet 11 inches tall. On her eleventh birthday she measured 4 feet 9 inches tall. How much had she grown?

67. Lisa has 5 packages to mail. If each package weighs 3 lbs. 6 oz., what is the total weight of the packages?

In problems 68 – 71, find the perimeter or circumference.

68.
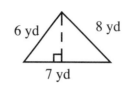

69.
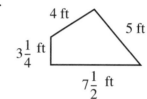

70.

71.
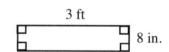

In problems 72 – 75, find the area.

72.
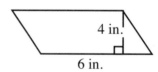

73.
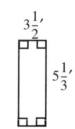

74.

75.

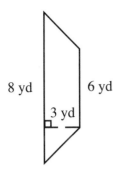

8 yd   6 yd

3 yd

In problems 76 – 80, find the volume.

76.

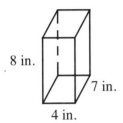

8 in.

7 in.

4 in.

77.

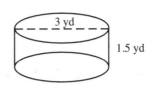

3 yd

1.5 yd

78.

10 in.

3 in.

79.

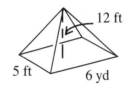

12 ft

5 ft   6 yd

80.

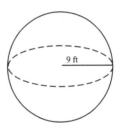

9 ft

 Check Yourself

1. $12 \text{yd} \cdot \dfrac{3 \text{ ft}}{1 \text{ yd}}$

    36 ft

    Multiply by a form of 1.
    Put the unit to be eliminated in the denominator.

    Perform the indicated operation. **(Conversion in the English system)**

2. $3 \text{ lb} \cdot \dfrac{16 \text{ oz}}{1 \text{ lb}}$

    48 oz **(Conversion in the English system)**

3. $6 \text{ pt} \cdot \dfrac{2 \text{ c}}{1 \text{ pt}}$

    12 c **(Conversion in the English system)**

4. $25 \text{ ft} \cdot \dfrac{1 \text{ yd}}{3 \text{ ft}}$

    $\dfrac{25}{3} = 8\dfrac{1}{3}$ yd **(Conversion in the English system)**

5. $9 \text{ pt} \cdot \dfrac{1 \text{ qt}}{2 \text{ pt}}$

$\dfrac{9}{2} = 4\dfrac{1}{2}$ qt **(Conversion in the English system)**

6. $2\dfrac{2}{3}$ yd   Change to an improper fraction first.

$\dfrac{8}{3}$ yd $\cdot \dfrac{3 \text{ ft}}{\text{yd}}$

$\dfrac{8}{3} \cdot \dfrac{3}{1} = 8$ ft   **(Conversion in the English system)**

7. $40 \text{ sec} \cdot \dfrac{1 \text{ min}}{60 \text{ sec}}$

$\dfrac{40}{60} = \dfrac{2}{3}$ min. **(Conversion in the English system)**

8. $90 \text{ oz} \cdot \dfrac{1 \text{ lb}}{16 \text{ oz}}$

$\dfrac{90}{16} = 5\dfrac{5}{8}$ lb **(Conversion in the English system)**

9. $7 \text{ wk} \cdot \dfrac{7 \text{ days}}{1 \text{ week}}$

49 days **(Conversion in the English system)**

10. $3.5 \text{ qt} \cdot \dfrac{2 \text{ pt}}{1 \text{ qt}} \cdot \dfrac{2 \text{ cups}}{1 \text{ pt}}$ Unless a direct conversion from quarts to cups is known, the problem must be done in two steps, converting quarts to pints and then pints to cups.

$(3.5)(2)(2) = 14$ cups **(Conversion in the English system)**

11. $3\dfrac{1}{2}$ hr

$\dfrac{7}{2}$ hr $\cdot \dfrac{60 \text{ min}}{1 \text{ hr}}$

210 min. **(Conversion in the English system)**

12. $20 \text{ cups} \cdot \dfrac{1 \text{ pt}}{2 \text{ cups}} \cdot \dfrac{1 \text{ qt}}{2 \text{ pt}} \cdot \dfrac{1 \text{ gal}}{4 \text{ qt}}$

$\dfrac{20}{2 \cdot 2 \cdot 4} = \dfrac{5}{4} = 1\dfrac{1}{4}$ gal **(Conversion in the English system)**

13. $\dfrac{3}{4}$ lb $\cdot \dfrac{16 \text{ oz}}{1 \text{ lb}}$

$\dfrac{3}{4} \cdot \dfrac{16}{1} = 12$ oz (**Conversion in the English system**)

14. 5.9 qt $\cdot \dfrac{2 \text{ pt}}{1 \text{ qt}}$

$(5.9)(2) = 11.8$ pt (**Conversion in the English system**)

15. 150 in $\cdot \dfrac{1 \text{ ft}}{12 \text{ in}}$

$\dfrac{150}{12} = 12\dfrac{1}{2}$ ft (**Conversion in the English system**)

16. 4.9 tons $\cdot \dfrac{2000 \text{ lb}}{1 \text{ ton}}$

$(4.9)(2000) = 9800$ lb (**Conversion in the English system**)

17. 10 hr $\cdot \dfrac{1 \text{ day}}{24 \text{ hr}}$

$\dfrac{10}{24} = \dfrac{5}{12}$ day (**Conversion in the English system**)

18. 176 oz $\cdot \dfrac{1 \text{ lb}}{16 \text{ oz}}$

$\dfrac{176}{16} = 11$ lb (**Conversion in the English system**)

19. $2\dfrac{3}{4}$ gal $\cdot \dfrac{4 \text{ qt}}{1 \text{ gal}} \cdot \dfrac{2 \text{ pt}}{1 \text{ qt}}$

$\dfrac{11}{4} \cdot \dfrac{4}{1} \cdot \dfrac{2}{1} = \dfrac{22}{1} = 22$ pt (**Conversion in the English system**)

20. 3.8 mi $\cdot \dfrac{5280 \text{ ft}}{1 \text{ mi}}$

$(3.8)(5280) = 20,064$ ft (**Conversion in the English system**)

21. 0.6 mg   Shift three places left. (**Conversion in the metric system**)

22. 0.00675 km   Shift five places left. (**Conversion in the metric system**)

23. 5000 cm   Shift four places right. (**Conversion in the metric system**)

24. 6090 dg   Shift one place right – add one zero. (**Conversion in the metric system**)

25. 0.0006 g   Shift three places left – add three zeros. (**Conversion in the metric system**)

26. 0.005 m   Shift three places left – add two zeros.
(Conversion in the metric system)

27. 0.00608 L   Shift three places left – add two zeros.
(Conversion in the metric system)

28. 9450 cL   Shift three places left – add one zero. **(Conversion in the metric system)**

29. 4000 m   Shift three places right – add three zeros.
(Conversion in the metric system)

30. 0.04000   Shift five places left – add one zero. **(Conversion in the metric system)**

31. $1\frac{5}{8} = 1.625$ cm   Change to a decimal.

    16.25 mm   Shift one place right. **(Conversion in the metric system)**

32. 6019 m   Shift three places right. **(Conversion in the metric system)**

33. 730 g   Shift three places right – add one zero.
(Conversion in the metric system)

34. 10.80 dg   Shift two places left. **(Conversion in the metric system)**

35. 420000 mL   Shift two places right – add three zeros.
(Conversion in the metric system)

36. 14920 L   Shift three places right – add one zero.
(Conversion in the metric system)

37. $2 \text{ L} \cdot \dfrac{1.06 \text{ qt}}{1 \text{ L}}$

    $(2)(1.06) = 2.12$ qt **(Conversions between English and metric units)**

38. $12 \text{ oz} \cdot \dfrac{1 \text{ lb}}{16 \text{ oz}} \cdot \dfrac{454 \text{ g}}{1 \text{ lb}}$ Change oz to lb, then to g.

    $\dfrac{(12)}{(1)} \dfrac{(1)}{(16)} \dfrac{(454)}{(1)} = 340\frac{1}{2}$ g **(Conversions between English and metric units)**

39. $15 \text{ kg} \cdot \dfrac{2.2 \text{ lb}}{1 \text{ kg}}$

    $(15)(2.2) = 33$ lb **(Conversions between English and metric units)**

40. $1 \text{ lb} \cdot \dfrac{454 \text{ g}}{1 \text{ lb}}$

    $(1)(454) = 454$ g **(Conversions between English and metric units)**

41. $2 \text{ km} \cdot \dfrac{0.6 \text{ mi}}{1 \text{ km}} \cdot \dfrac{5280 \text{ ft}}{1 \text{ mi}}$

    $(2)(0.6)(5280) = 6336$ ft **(Conversions between English and metric units)**

42. $600 \text{ m} \cdot \dfrac{39.4 \text{ in}}{1 \text{ m}} \cdot \dfrac{1 \text{ ft}}{12 \text{ in}} \cdot \dfrac{1 \text{ yd}}{3 \text{ ft}}$

$\dfrac{(600)\,(39.4)}{(12)\,(3)} = 656\frac{2}{3}$ yd **(Conversions between English and metric units)**

43. $6.36 \text{ L} \cdot \dfrac{1.06 \text{ qt}}{1 \text{ L}}$

$(6.3)(1.06) = 6.678$ qt **(Conversions between English and metric units)**

44. $20 \text{ yd} \cdot \dfrac{3 \text{ ft}}{1 \text{ yd}} \cdot \dfrac{12 \text{ in}}{1 \text{ ft}} \cdot \dfrac{1 \text{ m}}{39.4 \text{ in}}$

$\dfrac{(20)\,(3)\,(12)}{39.4} = \dfrac{720}{39.4} = 18.3$ m   Answer has been rounded to one decimal place. **(Conversions between English and metric units)**

45. $75 \text{ ft} \cdot \dfrac{12 \text{ in}}{1 \text{ ft}} \cdot \dfrac{1 \text{ m}}{39.4 \text{ in}}$

$\dfrac{(75)\,(12)\,(1)}{(1)\,(39.4)} = \dfrac{900}{39.4} = 22.8$ m   Answer has been rounded to one decimal place. **(Conversions between English and metric units)**

46. $8 \text{ lb} \cdot \dfrac{454 \text{ g}}{1 \text{ lb}}$

$(8)(454) = 3632$ g **(Conversions between English and metric units)**

47. $0.85 \text{ c} \cdot \dfrac{1 \text{ pt}}{2 \text{ c}} \cdot \dfrac{1 \text{ qt}}{2 \text{ pt}} \cdot \dfrac{946 \text{ mL}}{1 \text{ qt}}$

$\dfrac{(0.85)\,(1)\,(1)\,(946)}{(2)\,(2)\,(1)} = 201.025$ mL **(Conversions between English and metric units)**

48. $12 \text{ L} \cdot \dfrac{1.06 \text{ qt}}{1 \text{ L}}$

$(12)(1.06) = 12.72$ qt **(Conversions between English and metric units)**

49. $27.5 \text{ lb} \cdot \dfrac{1 \text{ kg}}{2.2 \text{ lb}}$

$\dfrac{(27.5)\,(1)}{2.2} = 12.5$ kg

50. $20 \text{ gal} \cdot \dfrac{4 \text{ qt}}{1 \text{ gal}} \cdot \dfrac{1 \text{ L}}{1.06 \text{ qt}}$ **(Conversions between English and metric units)**

$\dfrac{(20)\,(4)\,(1)}{(1)\,(1.06)} = 75.5$ L   Answer has been rounded to one decimal place. **(Conversions between English and metric units)**

51.　　$15 \text{ mi} \cdot \dfrac{1.61 \text{ km}}{1 \text{ mi}}$

$(15)(1.61) = 24.15$ km **(Conversions between English and metric units)**

52.　　6 ft　8 in.
　　　　　$\underline{\quad 2\quad}$
　　　　　12 ft　16 in.
　　　　　12 ft + 1 ft + 4 in.　　16 in. = 12 in. + 4 in.
　　　　　13 ft　4 in.
　　　　　　　　　　　　　　　= 1 ft + 4 in.

(Operations using English units)

53.　　　5 lb　7 oz　　　　　　4 lb + 1 lb　7 oz = 4 lb + 16 oz + 7 oz
　　　$\underline{\pm\ 3 \text{ lb}\ 10 \text{ oz}}$

　　　　4 lb　23 oz
　　　$\underline{\pm\ 3 \text{ lb}\ 10 \text{ oz}}$
　　　　1 lb　13 oz

(Operations using English units)

54.　　12 yd　　　11 in.　　　18 in. = 12 in. + 6 in. = 1 ft 6 in.
　　　$\underline{+15 \text{ yd}\ 2 \text{ ft}\ 7 \text{ in.}}$
　　　　27 yd　2 ft　18 in.
　　　　27 yd　2 ft + 1 ft 6 in.
　　　　27 yd　3 ft　6 in.
　　　　27 yd　1 yd　6 in.
　　　　28 yd　6 in.　　　　　3 ft = 1 yd

(Operations using English units)

55.　　　　$\overline{\text{2 min　8 sec}}$
　　　8$\overline{)\,}$17 min　4 sec
　　　　$\underline{16 \text{ min}}$
　　　　　1 min $\Rightarrow$ $\underline{60 \text{ sec}}$
　　　　　　　　64 sec
　　　　　　$\underline{64 \text{ sec}}$
　　　　　　　　0

(Operations using English units)

56.　　8 yd　　　　　　7 yd　3 ft　　　　　　7 yd　2 ft 12 in.
　　$\underline{-\quad\quad 5 \text{ ft}\ 3 \text{ in}}$　　$\underline{-\quad\quad 5 \text{ ft}\ 3 \text{ in}}$　　$\underline{-\quad\quad 5 \text{ ft}\ 3 \text{ in.}}$

　　　6 yd　5 ft　12 in.
　　$\underline{\pm\quad\quad 5 \text{ ft}\ 3 \text{ in.}}$
　　　6 yd　0 ft　9 in.

6 yd　9 in. **(Operations using English units)**

57.
$$\begin{array}{rr} 2 \text{ gal} & 3 \text{ qt} \\ +8 \text{ gal} & 3 \text{ qt} \\ \hline 10 \text{ gal} & 6 \text{ qt} \end{array}$$
    6 qt = 4 qt + 2 qt $\Rightarrow$ 1 gal + 2 qt

11 gal   2 qt
(Operations using English units)

58.
$$\begin{array}{r} 2 \text{ lb} \quad\;\; 14 \text{ oz} \\ 6\overline{)17 \text{ lb} \quad\;\; 4 \text{ oz}} \\ \underline{12 \text{ lb}} \quad\quad\quad \\ 5 \text{ lb} \Rightarrow \underline{80 \text{ oz}} \\ 84 \text{ oz} \\ \underline{84 \text{ oz}} \\ 0 \end{array}$$

(Operations using English units)

59.
$$\begin{array}{rr} 8 \text{ gal} & 3 \text{ qt} \\ \times & 4 \\ \hline 32 \text{ gal} & 12 \text{ qt} \end{array}$$
32 gal + 3 gal   12 qt = 4 qt + 4 qt + 4 qt
35 gal   1 gal + 1 gal + 1 gal
                                          3 gal

(Operations using English units)

60.
$$\begin{array}{r} 2 \text{ hr} \quad\;\; 15 \text{ min} \\ 5\overline{)11 \text{ hr} \quad\;\; 15 \text{ min}} \\ \underline{10 \text{ hr}} \quad\quad\quad \\ 1 \text{ hr} \Rightarrow \underline{60 \text{ min}} \\ 75 \text{ min} \\ \underline{75} \\ 0 \end{array}$$

(Word problems using English units)

61.
$$\begin{array}{r} 50 \\ \times\; 2 \\ \hline 100 \text{ ft} \end{array}$$
    $100 \text{ ft} \cdot \dfrac{1 \text{ yd}}{3 \text{ ft}}$

    $\dfrac{(100)\,(1)}{3} = 33\frac{1}{3} \text{ yd}$

He will have enough fencing material. **(Word problems using English units)**

62.
$$\begin{array}{rr} 6 \text{ ft} & 8 \text{ in.} \\ 6 \text{ ft} & 8 \text{ in.} \\ +3 \text{ ft} & 2 \text{ in.} \\ \hline 15 \text{ ft} & 18 \text{ in.} \end{array}$$
16 ft, 6 inches
$$\begin{array}{rr} 17 \text{ ft} & 2 \text{ in.} \\ -16 \text{ ft} & 6 \text{ in.} \\ \hline & 8 \text{ in. remain} \end{array}$$
(Word problems using English units)

63. $8\frac{1}{2} \Rightarrow 8.5$ oz

$$\begin{array}{r} 8.5 \\ \times\ \ \ 8 \\ \hline 68.0\ \ \text{oz} \end{array}$$

68 oz $\cdot \dfrac{1\ \text{lb}}{16\ \text{oz}} = 4\frac{1}{4}$ lb **(Word problems using English units)**

64.
$$\begin{array}{ll} 3\ \text{hrs} & 15\ \text{min} \\ \times & 5 \\ \hline 15\ \text{hrs} & 75\ \text{min} \\ 16\ \text{hrs} & 15\ \text{min} \end{array}$$
   75 min $= 60$ min $+ 15$ min $= 1$ hr $+ 15$ min

$$\begin{array}{ll} 16\ \text{hrs} & 15\ \text{min} \\ +\ 7\ \text{hrs} & 30\ \text{min} \\ \hline 23\ \text{hrs} & 45\ \text{min each week} \end{array}$$

(Word problems using English units)

65. 3 ft $\cdot \dfrac{12\ \text{in.}}{1\ \text{ft}} = 36$ in.

$$\begin{array}{r} 1\ \text{ft} \quad\ \ 9\ \text{in.} \\ 5\overline{)8\ \text{ft} \quad\ \ 9\ \text{in.}} \\ \underline{5\phantom{\ \text{ft}}} \\ 3\ \text{ft} \Rightarrow 36\ \text{in.} \\ \underline{45\ \text{in.}} \\ \pm 45\ \text{in.} \\ \hline 0 \end{array}$$

Each piece should be 1 ft 9 in. **(Word problems using English units)**

66.
$$\begin{array}{ll} 4\ \text{ft} & 9\ \text{in.} \\ \pm\ 3\ \text{ft} & 11\ \text{in.} \end{array} \Rightarrow \begin{array}{ll} 3\ \text{ft} + 1\ \text{ft} + 9\ \text{in.} \\ \pm\ 3\ \text{ft} \quad\quad 11\ \text{in.} \end{array} \Rightarrow \begin{array}{l} 3\ \text{ft} + 12\ \text{in.} + 9\ \text{in.} \end{array} \Rightarrow \begin{array}{ll} 3\ \text{ft} + 21\ \text{in.} \\ -\ 3\ \text{ft} \quad 11\ \text{in.} \\ \hline \quad\quad\quad 10\ \text{in.} \end{array}$$

She has grown 10 in. **(Word problems using English units)**

67.
$$\begin{array}{ll} 3\ \text{lb} & 6\ \text{oz} \\ \times & 5 \\ \hline 15\ \text{lb} & 30\ \text{oz} \\ 15\ \text{lb} + 1\ \text{lb} + 14\ \text{oz} \\ 16\ \text{lb} \quad 14\ \text{oz total weight} \end{array}$$
   30 oz $= 16$ oz $+ 14$ oz
   1 lb $+ 14$ oz

(Word problems using English units)

68. $P = 6$ yd $+ 8$ yd $+ 7$ yd

   21 yd (Perimeter)

69. $P = 4 \text{ ft} + 3 \text{ ft} + 5\frac{1}{4} \text{ ft} + 7\frac{1}{2} \text{ ft}$ Remember to find a common denominator for the fractions.

$19\frac{3}{4}$ ft **(Perimeter)**

70. $C = 2\pi r$

$= 2(3.14)(2.25)$

$= 14.13 \text{ ft}$ **(Circumference)**

71. $C = 3 \text{ ft} + 3 \text{ ft} + 8 \text{ in.} + 8 \text{ in.}$

$36 \text{ in.} + 36 \text{ in.} + 8 \text{ in.} + 8 \text{ in.}$ $3 \text{ ft} \cdot \dfrac{12 \text{ in.}}{1 \text{ ft}} = 36 \text{ in.}$

88 in **(Perimeter)**

72. $A = b \cdot h$

$6 \cdot 4$

24 sq in. **(Area)**

73. $A = l \cdot w$

$= \left(3\frac{1}{2} \right)\left(5\frac{1}{3} \right)$

$= \dfrac{7}{2} \cdot \dfrac{16}{3}$

$= \dfrac{56}{3} = 18\frac{2}{3} \text{ sq ft}$ **(Area)**

74. $A = \pi r^2$

$d = 8 \text{ yd}$

$r = 4 \text{ yd}$

$A = (3.14)\,(4)^2$

$= (3.14)(16)$

$= 50.24 \text{ sq yd}$ **(Area)**

75. $A = \dfrac{1}{2}(h)\,(b_1 + b_2)$

$= \dfrac{1}{2}(3)\,(8 + 6)$

$$= \frac{1}{2}(3)(14)$$

$$= 21 \text{ sq yd } (\textbf{Area})$$

76. $V = l \cdot w \cdot h$

$$= (7)(4)(8)$$

$$= 224 \text{ cu in. } (\textbf{Volume})$$

77. $V = \pi r^2 h$ $d = 3 \text{ yd}$

$$= (3.14)(1.5)^2(1.5) \qquad r = 1.5 \text{ yd}$$

$$= 10.5975 \text{ cu yd } (\textbf{Volume})$$

78. $V = \frac{1}{3}\pi r^2 h$

$$= \frac{1}{3}(3.14)(3)^2(10)$$

$$= \frac{1}{3}(3.14)(9)(10)$$

$$= 94.2 \text{ cu in. } (\textbf{Volume})$$

79. $V = \frac{1}{3}(l \cdot w \cdot h)$ $2 \text{ yd} = 6 \text{ ft}$

$$= \frac{1}{3}(6 \cdot 5 \cdot 12)$$

$$= \frac{1}{3}(360)$$

$$= 120 \text{ cu ft } (\textbf{Volume})$$

80. $V = \frac{4}{3}\pi r^3$

$$= \frac{4}{3}(3.14)(9)^3$$

$$= \frac{4}{3}(3.14)(729)$$

$$= 4(3.14)(243)$$

$$= 3052.08 \text{ cu ft } (\textbf{Volume})$$

Grade Yourself

Circle the question numbers that you had incorrect. Then indicate the number of questions you missed. If you answered more than three questions incorrectly, you need to focus on that topic. (If a topic has less than three questions and you had at least one wrong, we suggest you study that topic also. Read your textbook or a review book, or ask your teacher for help.)

Subject: *Measurement and Geometry*

Topic	Question Numbers	Number Incorrect
Conversion in the English system	1, 2, 3, 4, 5, 6, 7, 8, 9, 10, 11, 12, 13, 14, 15, 16, 17, 18, 19, 20	
Conversion in the metric system	21, 22, 23, 24, 25, 26, 27, 28, 29, 30, 31, 32, 33, 34, 35, 36	
Conversions between English and metric units	37, 38, 39, 40, 41, 42, 43, 44, 45, 46, 47, 48, 49, 50, 51	
Operations using English units	52, 53, 54, 55, 56, 57, 58, 59	
Word problems using English units	60, 61, 62, 63, 64, 65, 66, 67	
Perimeter	68, 69, 71	
Circumference	70	
Area	72, 73, 74, 75	
Volume	76, 77, 78, 79, 80	

Signed Numbers

 ## Brief Yourself

This chapter will review the rules for working with signed numbers. These rules will be applied to some of the topics from previous chapters such as fractions, decimals, exponents, and order of operations.

Negative Numbers are used to indicate quantities less than 0. These numbers are positioned to the left of 0 on a number line.

−4 is read "negative four"

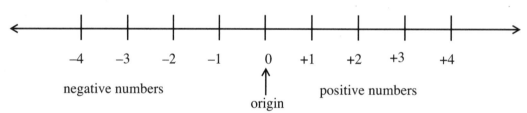

negative numbers   origin   positive numbers

A movement to the left on a number line is said to be in a negative direction. As you move to the left, the value of the numbers gets smaller. A movement to the right on a number line is said to be in a positive direction. As you move to the right, the value of the numbers gets larger.

When comparing the values of two numbers on a number line, the number on the left is always smaller than the number on the right.

Opposites — the two numbers that are the same distance from 0 with one in the negative direction and one in the positive direction.

6 and −6 are considered to be opposites

−(−8) can be read "The opposite of negative eight." Its value is 8.

Integers are the set of whole numbers and their opposites. $\{\ldots -3, -2, -1, 0, 1, 2, 3, \ldots\}$

Absolute Value of a number is the distance of that number from 0. Two vertical lines are used to denote absolute value.

Symbol	English	Mathematics
|6|	absolute value of 6 "How far from 0 is 6?"	$\|6\| = 6$
|−8|	absolute value of negative 8 "How far from 0 is −8?"	$\|-8\| = 8$

Symbol	*English*	*Mathematics*
$-\lvert-4\rvert$	the opposite of the absolute value of negative 4. "How far from 0 is -4?" What is the opposite of this number?	$-\lvert-4\rvert = -(4) = -4$

Absolute value measures the distance from 0, not the direction. Therefore, the absolute value of a number is always positive.

Addition

Same sign numbers:

— Add the numbers.

— Keep the same sign.

Different sign numbers:

— Subtract the smaller absolute value from the larger absolute value.

— Keep the sign from the number with the larger absolute value.

$$-10 + (-3) = -13 \qquad\qquad 7 + (-13) = -6$$

$$-6 + 2 = -4 \qquad\qquad -2 + (-5) = -7$$

Subtraction is the same as addition of the opposite:

— Do not change the number in front of the minus sign.

— Change subtraction to addition.

— Change the number after the subtraction sign to its opposite.

— Follow the rules for addition.

Examples:

$7 - 1$ becomes $7 + (-1) = 6$

$3 - (-2)$ becomes $3 + 2 = 5$

$-6 - 5$ becomes $-6 + (-5) = -11$

$-2 - (-7)$ becomes $-2 + 7 = 5$

$-3 + (-4) - 6 - (-2)$

— Read the problem first.

— Anytime you say the word *minus*, rewrite as shown above.

$-3 + (-4) - 6 - (-2)$ becomes $-3 + (-4) + (-6) + 2$

$$= -13 + 2 \qquad \text{Follow the rules for addition.}$$

$$= -11$$

Multiplication

Two numbers with the same sign:

— Multiply the numbers.

— Answer is positive.

Two numbers with different signs:

— Multiply the numbers.

— Answer is negative.

Examples:

$(-3) \, (-4) \; = \; 12$ $(-6) \, (7) \; = \; -42$

$(8) \, (9) \; = \; 72$ $(4) \, (-2) \; = \; -8$

Division

Two numbers with the same sign:

— Divide the numbers.

— Answer is positive.

Two numbers with different signs:

— Divide the numbers.

— Answer is negative.

Exponents: Use the explanation of exponents from Chapter 1, along with the explanation of signed numbers.

$$(-2)^4 \; = \; (-2) \, (-2) \, (-2) \, (-2)$$

$$= 4 \, (-2) \, (-2)$$

$$= (-8) \, (-2)$$

$$= 16$$

Note the differences between $(-3)^2, \, -(3)^2, \, -3^2$

$(-3)^2$ means $(-3) \, (-3) \; = \; 9$

$-(3)^2$ means the opposite of 3^2 $-(3) \, (3) \; = \; -9$

-3^2 means the opposite of 3^2 $-(3) \, (3) \; = \; -9$

Order of operations — Combine the rules of order of operations in Chapter 1 with signed number rules.

Test Yourself

1. Insert < or > between each pair of numbers so that the resulting statement is true.

   a. -11 -2

   b. 0 -1

   c. $-(-2)$ -4

   d. -11 $-(13)$

   e. $-\dfrac{1}{2}$ $-\dfrac{4}{5}$

   f. $-1\dfrac{5}{6}$ $1\dfrac{2}{3}$

   g. -4.5 -4.56

   h. -8.01 -8.1

2. Insert $<, >,$ or $=$ between each pair of numbers so that the resulting statement is true.

   a. $|-9|$ $|-15|$

   b. $|-23|$ $-(-23)$

   c. $-(-11)$ $-(-16)$

   d. $|4|$ $|0|$

   e. $|-5|$ $|0|$

   f. $-|-8|$ $-(-8)$

   g. $-|-4|$ $-(4)$

   h. $-|-6|$ $-|6|$

Add each of the following.

3. $-7 + 12$

4. $-6 + (-3)$

5. $9 + (-17)$

6. $-43 + (-9)$

7. $-68 + (-27)$

8. $34 + (-11) + (-13) + 128$

9. $13 + 19 + (-11)$

10. $-89 + (-42 + 37 + 19)$

11. $[8 + (-5)] + [3 + (-11)]$

12. $(-23 + 16) + (-17 + 4)$

13. $\dfrac{5}{12} + \left(-\dfrac{3}{8}\right)$

14. $-\dfrac{1}{6} + \dfrac{1}{2}$

15. $22\dfrac{4}{9} + \left(-13\dfrac{1}{3}\right)$

16. $-\dfrac{1}{20} + \dfrac{8}{30}$

17. $-4 + 2\dfrac{1}{8}$

18. $(-6.54) + (-3.87)$

19. $(-4.5) + 8.67$

20. $(-17.4) + (2.687)$

21. $(-3.2) + (4.92) + (-7.91) + (4.6)$

22. $(3.95) + (-15.5) - (13.16) + (29.4)$

Subtract each of the following.

23. $-7 - 2$

24. $-20 - (-40)$

25. $9 - (-6)$

26. $25 - (-25)$

27. $2 - 42$

28. $-10 - 6 - (-7)$

29. $-43 - (-24) - 10 - (-16)$

30. $47 - 21 - (-12)$

31. $3 - (-2 - 6)$

32. $-2 - (6 - 15)$

33. $\frac{1}{5} - \left(-\frac{7}{8}\right)$

34. $-\frac{3}{4} - \frac{1}{14}$

35. $-2\frac{1}{5} - 3\frac{1}{4}$

36. $-6\frac{1}{3} - \left(-1\frac{4}{5}\right)$

37. $-\frac{5}{12} - \left(-\frac{3}{8}\right)$

38. $(-7.32) - (4.61)$

39. $(-19.67) - (-21.94)$

40. $-9.2 - (11.57)$

41. $(-10.42) - (-10.9) - (9.4) - (-31.6)$

42. $(-7.6) - (2.89) - (-16.71) - (10.5)$

Multiply each of the following.

43. $(7)\,(-6)$

44. $(-5)\,(-8)$

45. $(-2)\,(3)\,(-7)$

46. $(-6)\,(-4)\,(-2)$

47. $(-4)^3$

48. $-(5)^2$

49. $(-10)\,(-5)\,(0)$

50. $(-1)^4$

51. $[\,(-2)\,(3)\,]^2$

52. $(-2)\,(3)\,(-2)\,(-3)$

53. $\left(-\frac{3}{5}\right)\left(\frac{5}{2}\right)$

54. $\left(-2\frac{5}{8}\right)\left(-1\frac{1}{2}\right)$

55. $\left(-\frac{1}{2}\right)^2\left(\frac{3}{5}\right)^3$

56. $\left(\frac{27}{40}\right)\left(-\frac{8}{5}\right)$

57. $\left(-3\frac{1}{5}\right)\left(-5\frac{1}{6}\right)\left(-1\frac{1}{8}\right)$

58. $(-0.4)\,(-3.2)$

59. $(0.3)\,(-0.2)$

60. $(1.2)\,(-1.3)\,(0.4)$

61. $(61.3)\,(-0.425)$

62. $(-4.2)\,(-1.6)\,(-0.25)$

Divide each of the following.

63. $12 \div (-4)$

64. $-15 \div (-3)$

65. $0 \div (-14)$

66. $-18 \div (-18)$

67. $-20 \div 0$

68. $-80 \div (-10) \div 2$

69. $-500 \div (-50 \div 10)$

70. $(-500 \div -50) \div 10$

71. $-80 \div 2 \div 10$

72. $-475 \div 25$

73. $\left(-\frac{3}{4} \div 2\frac{1}{2}\right) \div 3$

74. $\frac{11}{19} \div \left(-\frac{15}{19}\right)$

75. $\left(-\frac{7}{9}\right) \div \left(\frac{9}{5}\right)$

76. $-\frac{14}{15} \div \left(-\frac{5}{7}\right)$

77. $6\frac{1}{4} \div \left(-3\frac{3}{4}\right)$

78. $8.4 \div (-0.4)$

79. $-0.45 \div (1.2)$

80. $-77.6 \div (-8)$

81. $28.8 \div (-0.6)$

82. $-56.602 \div 1.4$

Perform the indicated operations.

83. $8(-7) + 6(-5)$

84. $19 - 5(-3) + 3$

85. $8[(6-13)-11]$

86. $(12-19) \div 7$

87. $8^2 - (5-3)^4$

88. $6[9-(3-4)]$

89. $-3 + 2(3+4)$

90. $-5 - [13-7] - (-5+9)$

91. $-[-2-(-4-6)]$

92. $-9 - [6 + 2(-5+8)]$

93. $12 - 2(3-5) + \dfrac{14+6}{3^2+1}$

94. $\dfrac{7(-2)-6}{-10}$

95. $\dfrac{6(-7)+3(-2)}{20-4}$

96. $-[1-3(2-7)^2]$

97. $22 - [(-6+2)-8] - 12$

Check Yourself

1.   a.   <   When comparing two negative numbers, picture them on a number line. The number on the left is smaller.

      b.   >   0 is always greater than a negative number.

      c.   >   A positive number is always greater than a negative number.

      d.   >

      e.   >

      f.   <

      g.   >

      h.   >

     (Integers — inequalities)

2.   a.   <   Both are positive.

      b.   =   Both are positive.

      c.   <   Both are positive.

d. >

e. >

f. <

g. =

h. =

(Integers — absolute value)

3. 5   Subtract, keep sign from 12. **(Addition of integers)**

4. –9   Add. **(Addition of integers)**

5. –8   Subtract, keep sign from –17 . **(Addition of integers)**

6. –52   Add. **(Addition of integers)**

7. –95   Add. **(Addition of integers)**

8. 162 + (–24)   Add positives, add negatives.

    = 138   Subtract, keep sign from 162. **(Addition of integers)**

9. 32 + (–11)   Add positives.

    = 21   Subtract, keep sign from 32. **(Addition of integers)**

10. – 131 + 56   Add positives, add negatives.

    = –75   Subtract, keep sign from –131 . **(Addition of integers)**

11. 3 + (–8)   Subtract quantities in [ ].

    = –5   Subtract, keep sign from –8 . **(Addition of integers)**

12. –7 + –13

    = –20   **(Addition of integers)**

13. $\dfrac{5}{12} + \left(-\dfrac{3}{8}\right)$

    $= \dfrac{10}{24} + \dfrac{-9}{24}$ Find LCD.

    $= \dfrac{1}{24}$ Subtract numerators, keep sign from 10. **(Addition of integers)**

14. $\dfrac{-1}{6} + \dfrac{1}{2}$

    $= \dfrac{-1}{6} + \dfrac{3}{6}$ Find LCD.

$$= \frac{2}{6}$$ Subtract numerators, keep sign from 3.

$$= \frac{1}{3}$$ Reduce to lowest terms. **(Addition of integers)**

15. $22\frac{4}{9} + -13\frac{1}{3}$

$$= \frac{202}{9} + \frac{-40}{3}$$ Change to improper fractions.

$$= \frac{202}{9} + \frac{-120}{9}$$ Find LCD.

$$= \frac{82}{9}$$ Subtract numerators, keep sign from 202.

$$= 9\frac{1}{9}$$ Change to mixed number. **(Addition of integers)**

16. $\frac{-1}{20} + \frac{8}{30}$

$$= \frac{-3}{60} + \frac{16}{60}$$ Find LCD.

$$= \frac{13}{60}$$ Subtract numerators, keep sign from 16. **(Addition of integers)**

17. $-4 + 2\frac{1}{8}$

$$= \frac{-4}{1} + \frac{17}{8}$$ Change to improper fractions.

$$= \frac{-32}{8} + \frac{17}{8}$$ Find LCD.

$$= \frac{-15}{8}$$ Subtract numerators, keep sign from 17.

$$= -1\frac{7}{8}$$ Change to mixed number. **(Addition of integers)**

18. $(-6.54) + (-3.87)$

$$= -10.41$$ Add, keep same sign. **(Addition of integers)**

19. $(-4.5) + (8.67)$

$$= 4.17$$ Subtract, keep sign from 8.67. **(Addition of integers)**

20. $(-17.4) + (2.687)$

       $= -14.713$ Subtract, keep sign from 17.4. (**Addition of integers**)

21. $(-3.2) + (4.92) + (-7.91) + (4.6)$

       $= -11.11 + 9.52$ Add like signed numbers, keep same sign.

       $= -1.59$ Subtract, keep sign from 11.11. (**Addition of integers**)

22. $(3.95) + (-15.5) + (-13.16) + (29.4)$

       $= 33.35 + (-28.66)$ Add like signed numbers, keep same sign.

       $= 4.69$ Subtract, keep sign from 33.35. (**Addition of integers**)

23. $-7 - 2$

       $= -7 + (-2)$ Rewrite as addition of the opposite.

       $= -9$ Add, keep same sign. (**Subtraction of integers**)

24. $-20 - (-40)$

       $= -20 + (40)$ Rewrite as addition of the opposite.

       $= 20$ Subtract, keep sign from 40. (**Subtraction of integers**)

25. $9 - (-6)$

       $= 9 + 6$ Rewrite as addition of the opposite.

       $= 15$ Add, keep same sign. (**Subtraction of integers**)

26. $25 - (-25)$

       $= 25 + 25$ Rewrite as addition of the opposite.

       $= 50$ Add, keep same sign. (**Subtraction of integers**)

27. $2 - 42$

       $= 2 + (-42)$ Rewrite as addition of the opposite.

       $= -40$ Subtract, keep sign from 42. (**Subtraction of integers**)

28. $-10 - 6 - (-7)$

       $= -10 + (-6) + 7$ Rewrite all subtractions as additions of the opposite.

       $= -16 + 7$ Add like signed numbers, keep same sign.

       $= -9$ Subtract, keep sign from 16. (**Subtraction of integers**)

29. $-43 - (-24) - 10 - (-16)$

       $= -43 + 24 + (-10) + 16$ Rewrite all subtractions as additions of the opposite.

$= -53 + 40$ Add liked signed numbers, keep same sign.

$= -13$ Subtract, keep sign from 53. (**Subtraction of integers**)

30. $47 - 21 - (-12)$

     $= 47 + (-21) + 12$ Rewrite all subtraction as addition of the opposite.

     $= 59 + (-21)$ Add like signed numbers, keep same sign.

     $= 38$ Subtract, keep sign from 59. (**Subtraction of integers**)

31. $3 - (-2 - 6)$

     $= 3 - (-2 + (-6))$ Rewrite all subtractions as additions of the opposite.

     $= 3 - (-8)$ Add like signed numbers inside parentheses.

     $= 3 + 8$ Rewrite all subtractions as additions of the opposite.

     $= 11$ Add, keep same sign. (**Subtraction of integers**)

32. $-2 - (6 - 15)$

     $= -2 - (6 + (-15))$ Rewrite all subtractions as additions of the opposite.

     $= -2 - (-9)$ Subtract signed numbers inside parentheses.

     $= -2 + (9)$ Rewrite all subtractions as additions of the opposite.

     $= 7$ Add, keep sign of 9. (**Subtraction of integers**)

33. $\dfrac{1}{5} - \left(-\dfrac{7}{8}\right)$

     $= \dfrac{1}{5} + \dfrac{7}{8}$ Rewrite subtraction as addition of opposite.

     $= \dfrac{8}{40} + \dfrac{35}{40}$ Find LCD.

     $= \dfrac{43}{40}$ Add numerators, keep same sign.

     $= 1\dfrac{3}{40}$ Change to mixed number. (**Subtraction of integers**)

34. $-\dfrac{3}{4} - \dfrac{1}{14}$

     $= -\dfrac{3}{4} + \left(-\dfrac{1}{14}\right)$ Rewrite subtraction as addition of the opposite.

$$= -\frac{21}{28} + \left(-\frac{2}{28}\right)$$ Find LCD.

$$= -\frac{23}{28}$$ Add like signed numerators, keep same sign. **(Subtraction of integers)**

35. $-2\frac{1}{5} - 3\frac{1}{4}$

$$= -\frac{11}{5} - \frac{13}{4}$$ Change to improper fractions.

$$= -\frac{11}{5} + \left(-\frac{13}{4}\right)$$ Rewrite subtraction as addition of the opposite.

$$= -\frac{44}{20} + \left(-\frac{65}{20}\right)$$ Find LCD.

$$= \frac{-109}{20}$$ Change to mixed number.

$$= -5\frac{9}{20}$$ **(Subtraction of integers)**

36. $-6\frac{1}{3} - \left(-1\frac{4}{5}\right)$

$$= -\frac{19}{3} - \left(-\frac{9}{5}\right)$$ Change to improper fractions.

$$= -\frac{19}{3} + \frac{9}{5}$$ Rewrite subtraction as addition of the opposite.

$$= -\frac{95}{15} + \frac{27}{15}$$ Find LCD.

$$= -\frac{68}{15}$$ Change to mixed number.

$$= -4\frac{8}{15}$$ **(Subtraction of integers)**

37. $-\frac{5}{12} - \left(-\frac{3}{8}\right)$

$$= -\frac{5}{12} + \frac{3}{8}$$ Rewrite subtraction as addition of the opposite.

$$= -\frac{10}{24} + \frac{9}{24}$$ Find LCD.

$$= -\frac{1}{24}$$ Add like signed numerators, keep same sign. **(Subtraction of integers)**

38. $-7.32 - 4.61$

$= (-7.32) + (-4.61)$ Rewrite subtraction as addition of the opposite.

$= -11.93$ Add, keep same sign. **(Subtraction of integers)**

39. $(-19.67) - (-21.94)$

$= (-19.67) + (21.94)$ Rewrite subtraction as addition of the opposite.

$= 2.27$ Subtract, keep sign from 21.94. **(Subtraction of integers)**

40. $-9.2 - (11.57)$

$= -9.2 + (-11.57)$ Rewrite subtraction as addition of the opposite.

$= -20.77$ Add, keep same sign. **(Subtraction of integers)**

41. $(-10.42) - (-10.9) - (9.4) - (-31.6)$

$= (-10.42) + (10.9) + (-9.4) + (31.6)$ Rewrite all subtractions as additions of the opposite.

$= -19.82 + 42.5$ Add like signed numbers, keep same sign.

$= 22.68$ Subtract, keep sign from 42.5. **(Subtraction of integers)**

42. $(-7.6) - (2.89) - (-16.71) - (10.5)$

$= (-7.6) + (-2.89) + (16.71) + (-10.5)$ Rewrite all subtraction as addition of the opposite.

$= 16.71 + (-20.99)$ Add like signed numbers, keep same sign.

$= -4.28$ Subtract, keep sign from 20.99. **(Subtraction of integers)**

43. $(7)\,(-6)$

$= -42$ **(Multiplication of integers)**

44. $(-5)\,(-8)$

$= 40$ **(Multiplication of integers)**

45. $(-2)\,(3)\,(-7)$ These can be multiplied in any order since multiplication is commutative.

$= (-6)\,(-7)$

$= 42$ **(Multiplication of integers)**

46. $(-6)\,(-4)\,(-2)$

   $= (24)\,(-2)$

   $= -48$ **(Multiplication of integers)**

47. $(-4)^{3}$

   $= (-4)\,(-4)\,(-4)$

   $= (16)\,(-4)$

   $= -64$ **(Multiplication of integers)**

48. $-(5)^{2}$

   $= -(5)\,(5)$

   $= -25$ **(Multiplication of integers)**

49. $(-10)\,(-5)\,(0)$

   $= 0$ **(Multiplication of integers)**

50. $(-1)^{4}$

   $= (-1)\,(-1)\,(-1)\,(-1)$

   $= 1\,(-1)\,(-1)$

   $= (-1)\,(-1)\ =\ 1$ **(Multiplication of integers)**

51. $[\,(-2)\,(3)\,]^{2}$

   $= [-6]^{2}$

   $= (-6)\,(-6)$

   $= 36$ **(Multiplication of integers)**

52. $(-2)\,(3)\,(-2)\,(-3)$

   $= (-6)\,(-2)\,(-3)$

   $= 12\,(-3)$

   $= -36$ **(Multiplication of integers)**

53. $\left(-\dfrac{3}{5}\right)\left(\dfrac{5}{2}\right)$

    $= -\dfrac{3}{2}$

    $= -1\dfrac{1}{2}$ **(Multiplication of integers)**

54. $\left(-2\dfrac{5}{8}\right)\left(-1\dfrac{1}{2}\right)$

    $= \left(-\dfrac{21}{8}\right)\left(-\dfrac{3}{2}\right)$

    $= \dfrac{63}{16}$

    $= 3\dfrac{15}{16}$ **(Multiplication of integers)**

55. $\left(-\dfrac{1}{2}\right)^{2}\left(\dfrac{3}{5}\right)^{3}$

    $= \left(-\dfrac{1}{2}\right)\left(-\dfrac{1}{2}\right)\left(\dfrac{3}{5}\right)\left(\dfrac{3}{5}\right)\left(\dfrac{3}{5}\right)$

    $= \left(\dfrac{1}{4}\right)\left(\dfrac{3}{5}\right)\left(\dfrac{3}{5}\right)\left(\dfrac{3}{5}\right)$

    $= \left(\dfrac{3}{20}\right)\left(\dfrac{3}{5}\right)\left(\dfrac{3}{5}\right)$

    $= \left(\dfrac{9}{100}\right)\left(\dfrac{3}{5}\right)$

    $= \dfrac{27}{500}$ **(Multiplication of integers)**

56. $\left(\dfrac{27}{40}\right)\left(-\dfrac{8}{5}\right)$

    $= -\dfrac{27}{25}$

    $= -1\dfrac{2}{25}$ **(Multiplication of integers)**

57. $\left(-\dfrac{16}{5}\right)\left(-\dfrac{31}{6}\right)\left(-\dfrac{9}{8}\right)$

    $= -18\dfrac{3}{5}$ **(Multiplication of integers)**

58. (−0.4) (−3.2)

= 1.28   **(Multiplication of integers)**

59. (0.3) (−0.2)

= −0.06   **(Multiplication of integers)**

60. (1.2) (−1.3) (0.4)

= −0.624   **(Multiplication of integers)**

61. (61.3) (−0.425)

= −26.0525   **(Multiplication of integers)**

62. (−4.2) (−1.6) (−0.25)

= −1.68   **(Multiplication of integers)**

63. $12 \div -4$

= −3   **(Division of integers)**

64. $-15 \div (-3)$

= 5   **(Division of integers)**

65. $0 \div (-14)$ 0 divided by any non-zero number is 0.

= 0   **(Division of integers)**

66. $-18 \div (-18)$

= 1   **(Division of integers)**

67. $-20 \div 0$ Division by 0 is an undefined operation.

undefined   **(Division of integers)**

68. $-80 \div (-10) \div 2$ Must complete the operations from left to right.

$= 8 \div 2$

= 4   **(Division of integers)**

69. $-500 \div (-50 \div 10)$ Evaluate the operation in the parentheses.

$= -500 \div -5$

= 100   **(Division of integers)**

70. $(-500 \div -50) \div 10$ Evaluate the operation in the parentheses.

$= 10 \div 10$

= 1   **(Division of integers)**

71. $-80 \div 2 \div 10$

     $= -40 \div 10$

     $= -4$ **(Division of integers)**

72. $-475 \div 25$

     $= -19$ **(Division of integers)**

73. $\left(-\dfrac{3}{4} \div 2\dfrac{1}{2} \right) \div 3$

     $= \left(-\dfrac{3}{4} \div \dfrac{5}{2} \right) \div 3$

     $= \left(-\dfrac{3}{4} \right)\left(\dfrac{2}{5} \right) \div 3$ Rewrite division as multiplication by the reciprocal.

     $= -\dfrac{3}{10} \div 3$

     $= \left(-\dfrac{3}{10} \right)\left(\dfrac{1}{3} \right)$ Rewrite division as multiplication by the reciprocal.

     $= -\dfrac{1}{10}$ **(Division of integers)**

74. $\dfrac{11}{19} \div \left(-\dfrac{15}{19} \right)$

     $= \dfrac{11}{19} \cdot -\dfrac{19}{15}$ Rewrite division as multiplication by the reciprocal.

     $= -\dfrac{11}{15}$ **(Division of integers)**

75. $-\dfrac{7}{9} \div \dfrac{9}{5}$

     $= -\dfrac{7}{9} \cdot \dfrac{5}{9}$ Rewrite division as multiplication by the reciprocal.

     $= -\dfrac{35}{81}$ **(Division of integers)**

76. $-\dfrac{14}{15} \div \left(-\dfrac{5}{7} \right)$

     $= -\dfrac{14}{15} \cdot -\dfrac{7}{5}$

$$= \frac{98}{75}$$

$$= 1\frac{23}{75} \qquad \textbf{(Division of integers)}$$

77. $\quad 6\frac{1}{4} \div -3\frac{3}{4}$

$$= \frac{25}{4} \div -\frac{15}{4} \qquad \text{Change to improper fractions.}$$

$$= \frac{25}{4} \cdot -\frac{4}{15} \qquad \text{Rewrite division as multiplication by the reciprocal.}$$

$$= -\frac{5}{3}$$

$$= -1\frac{2}{3} \qquad \text{Change to mixed number. } \textbf{(Division of integers)}$$

78. $\quad 8.4 \div -0.4$

$$= \quad -0.4\overline{)8.4}$$

$$= \quad \begin{array}{r} -21 \\ -4\overline{)84} \\ \underline{8} \\ 4 \\ \underline{4} \\ 0 \end{array} \qquad \begin{array}{l}\text{Shift decimal points one place to the right in} \\ \text{both the divisor and the dividend.}\end{array}$$

$8.4 \div 0.4 = -21 \qquad \textbf{(Division of integers)}$

79. $\quad -0.45 \div 1.2$

$$= \quad 1.2\overline{)-0.45}$$

$$= \quad \begin{array}{r} -.375 \\ 12\overline{)-04.500} \\ \underline{36} \\ 90 \\ \underline{84} \\ 60 \\ \underline{60} \\ 0 \end{array} \qquad \begin{array}{l}\text{Negative answer.} \\ \\ \text{Shift decimal points.} \\ \text{Add zero, continue to divide.} \\ \\ \text{Add zero, continue to divide.}\end{array}$$

(Division of integers)

80. $-77.6 \div (-8)$

$$= \begin{array}{r} 9.7 \\ -8\overline{)-77.6} \\ \underline{72} \\ 5\ 6 \\ \underline{5\ 6} \\ 0 \end{array}$$

Division of two negatives yields a positive answer.
Bring decimal point straight up into the quotient.

(Division of integers)

81. $28.8 \div (-0.6)$

$$= \quad -0.6\overline{)28.8}$$

$$= \begin{array}{r} -48 \\ -6\overline{)288} \\ \underline{24} \\ 48 \\ \underline{48} \\ 0 \end{array}$$

Negative answer.

Shift decimal points.

(Division of integers)

82. $-56.602 \div 1.4$

$$= \quad 1.4\overline{)-56.602}$$

$$= \begin{array}{r} -40.43 \\ 14\overline{)-566.02} \\ \underline{56} \\ 6 \\ \underline{0} \\ 6\ 0 \\ \underline{5\ 6} \\ 42 \\ \underline{42} \\ 0 \end{array}$$

Negative answer.

Shift decimal points.

(Division of integers)

83. $8\,(-7) + 6\,(-5)$

$= -56 + (-30)$

$= -86$ **(Integers — order of operations)**

84. $19 - 5\,(-3) + 3$

$= 19 + (-5)\,(-3) + 3$ Rewrite subtraction as addition of the opposite.

$= 19 + 15 + 3$

$= 37$ **(Integers — order of operations)**

85. $8 [(6 - 13) - 11]$

$= 8 [(6 + (-13)) + (-11)]$ Rewrite subtractions as additions of the opposite.

$= 8 [- 7 + (-11)]$ Complete the operation inside the parentheses.

$= 8 [-18]$

$= -144$ **(Integers — order of operations)**

86. $(12 - 19) \div 7$

$= (12 + (-19)) \div 7$ Rewrite subtraction as addition of the opposite.

$= -7 \div 7$

$= -1$ **(Integers — order of operations)**

87. $8^2 - (5 - 3)^4$ Complete operation inside the parentheses.

$= 8^2 - (2)^4$ Evaluate the exponents.

$= 64 - 16$

$= 48$ **(Integers — order of operations)**

88. $6 [9 - (3 - 4)]$

$= 6 [-(3 + (-4))]$

$= 6 [9 - (-1)]$

$= 6 [9 + 1]$

$= 6[10]$

$= 60$ **(Integers — order of operations)**

89. $- 3 + 2 (3 + 4)$

$= - 3 + 2 (7)$ Multiply before adding.

$= - 3 + 14$

$= 11$ **(Integers — order of operations)**

90. $- 5 - [13 - 7] - (-5 + 9)$

$= - 5 - [13 + (-7)] - (-5 + 9)$

$= - 5 - [6] - (4)$

$$= -5 + [-6] + (-4)$$

$$= -11 + (-4)$$

$$= -15 \qquad \text{(Integers — order of operations)}$$

91. $-[-2-(-4-6)]$

$$= -[-2-(-4+(-6))]$$

$$= -[-2-(-10)]$$

$$= -[-2+10]$$

$$= -[8]$$

$$= -8 \qquad \text{(Integers — order of operations)}$$

92. $-9-[6+2(-5+8)]$

$$= -9-[6+2(3)]$$

$$= -9-[6+6] \qquad \text{Multiply before adding.}$$

$$= -9-12$$

$$= -9+(-12)$$

$$= -21 \qquad \text{(Integers — order of operations)}$$

93. $12-2(3-5)+\dfrac{14+6}{3^2+1}$ Fraction bar acts like a grouping symbol.

$$= 12-2(3-5)+\frac{20}{9+1} \qquad \text{Complete the numerator, the denominator, then the fraction as a whole.}$$

$$= 12-2(3-5)+\frac{20}{10}$$

$$= 12-2(3-5)+2$$

$$= 12-2(-2)+2 \qquad \text{Multiply before adding.}$$

$$= 12+(-2)(-2)+2$$

$$= 12+4+2$$

$$= 16+2$$

$$= 18 \qquad \text{(Integers — order of operations)}$$

94. $\dfrac{7(-2) - 6}{-10}$

    $= \dfrac{-14 - 6}{-10}$

    $= \dfrac{-14 + (-6)}{-10}$

    $= \dfrac{-20}{-10}$

    $= 2$ **(Integers — order of operations)**

95. $\dfrac{6(-7) + 3(-2)}{20 - 4}$

    $= \dfrac{-42 + (-6)}{20 + (-4)}$

    $= \dfrac{-48}{16}$

    $= -3$ **(Integers — order of operations)**

96. $-[1 - 3(2 - 7)^2]$

    $= -[1 - 3(2 + (-7))^2]$

    $= -[1 - 3(-5)^2]$

    $= -[1 + (-3)(25)]$

    $= -[1 + (-75)]$

    $= -[-74]$

    $= 74$ **(Integers — order of operations)**

97. $22 - [(-6 + 2) - 8] - 12$

    $= 22 - [-4 - 8] - 12$

    $= 22 - [-4 + (-8)] - 12$

    $= 22 - [-12] - 12$

    $= 22 + 12 + (-12)$

    $= 34 + (-12)$

    $= 22$ **(Integers — order of operations)**

Grade Yourself

Circle the question numbers that you had incorrect. Then indicate the number of questions you missed. If you answered more than three questions incorrectly, you need to focus on that topic. (If a topic has less than three questions and you had at least one wrong, we suggest you study that topic also. Read your textbook or a review book, or ask your teacher for help.)

Subject: *Signed Numbers*

Topic	Question Numbers	Number Incorrect
Integers — inequalities	1	
Integers — absolute value	2	
Addition of integers	3, 4, 5, 6, 7, 8, 9, 10, 11, 12, 13, 14, 15, 16, 17, 18, 19, 20, 21, 22	
Subtraction of integers	23, 24, 25, 26, 27, 28, 29, 30, 31, 32, 33, 34, 35, 36, 37, 38, 39, 40, 41, 42	
Multiplication of integers	43, 44, 45, 46, 47, 48, 49, 50, 51, 52, 53, 54, 55, 56, 57, 58, 59, 60, 61, 62	
Division of integers	63, 64, 65, 66, 67, 68, 69, 70, 71, 72, 73, 74, 75, 76, 77, 78, 79, 80, 81, 82	
Integers — order of operations	83, 84, 85, 86, 87, 88, 89, 90, 91, 92, 93, 94, 95, 96, 97	

Introduction to Polynomials

8

 Brief Yourself

This chapter will review the evaluation of algebraic expressions and also provide an expanded presentation of equation solving. Algebraic word problems are also be discussed.

Constants are mathematical expressions whose values never change.

Examples: $4, -\frac{1}{2}, 6$

Variables are mathematical expressions whose values change from problem to problem.

Examples: x, y, a, A

Terms are numbers or the products of numbers and variables raised to powers.

Examples: $2, \frac{1}{2}x, -7x^2, y$

Numerical Coefficients are the number part of a term.

$3x$	coefficient $= 3$
x^2	coefficient $= 1$
$-x^3$	coefficient $= -1$
$\dfrac{x^2}{2}$	coefficient $= \dfrac{1}{2}$

Like Terms are terms that have the same variables raised to the same power - coefficients need not be the same

Examples: $3xy^3 \ -7xy^3$ and $6y^3x$ are all like terms.

$2x$ and $2y$ are not like terms.

Algebraic Expression are the sum or difference of terms.

$3x \qquad 4x^2 - 8x \qquad ab + c + 5$

To simplify an algebraic expression means to combine the like terms. To combine like terms, add or subtract the numerical coefficients and attach the common variables.

NOTE: In most cases, subtraction should be rewritten as addition of the opposite.

It is often necessary to use the commutative and/or the associative properties. Remember, these properties hold for addition and multiplication only. This is why subtraction is rewritten as addition.

Example:

$2y - 6 + 5y + 2$ becomes $2y + (-6) + 5y + 2$

$$= 2y + 5y + (-6) + 2$$

$$= 7y + (-4)$$

$$= 7y - 4$$

To evaluate an expression, replace the variables with the values that have been provided. Use the order of operation rules to perform the operations.

Recall the distributive property from Chapter 1.

$$a(b + c) = (a)(b) + (a)(c)$$

$$3(5x + 4) = 3(5x) + 3(4)$$

$$= 15x + 12$$

$$-(3x - 4)$$

$$= -1(3x - 4)$$

$$= -1(3x) - (-1)(4)$$

$$= -3x + 4$$

$$2(x - 6)$$

$$= 2(x) - 2(6)$$

$$= 2x - 12$$

Now that the introductory algebra topics have been reviewed, the rules for solving equations (see Chapter 1) can be expanded.

1. Rewrite any subtractions as additions of the opposite.

2. Use the distributive property to clear parentheses.

3. Collect any like terms that are on the same side of the equal sign.

4. Move like terms to the same side by adding the opposite of the term that is being moved to both sides.

5. Move the numerical coefficient to isolate the variable by dividing both sides of the equation by the coefficient or multiplying both sides of the equation by the reciprocal of the coefficient.

Example:

$$14 + 4(w - 5) = 6 - 2w$$

$$14 + 4w + (-20) = 6 + (-2w)$$ Distributive property

$$4w + (-6) = 6 + (-2w)$$ Combine like terms that are on the same side of the equal sign.

$$4w + 2w + (-6) = 6 + (-2w) + 2w$$ Combine like terms that are on the opposite side of the equal sign by adding the opposite of the term being moved to both sides.

$$6w + (-6) + 6 = 6 + 6$$

$$6w = 12$$

$$\frac{6w}{6} = \frac{12}{6}$$ Divide both sides by 6.

$$w = 2$$

NOTE: When an equation contains fractions, it can be solved by leaving the fractions in the equation and using the appropriate fraction rules. However, the fractions can be eliminated from the equation by multiplying each term of the equation by the LCD (least common denominator).

Example:

$$\frac{2}{3}(x + 4) = \frac{1}{2}x - \frac{5}{6}$$ Use the distributive property.

$$\frac{2}{3}x + \frac{8}{3} = \frac{1}{2}x + \left(-\frac{5}{6}\right)$$ Rewrite subtraction as addition of the opposite.

$$6 \cdot \frac{2}{3}x + 6 \cdot \frac{8}{3} = 6 \cdot \frac{1}{2}x + 6 \cdot \left(-\frac{5}{6}\right)$$ 6 is the LCD for these fractions.

$$4x + 16 = 3x + (-5)$$ All terms must be multiplied by the LCD.

$$4x + (-3x) + 16 = 3x + (-3x) + (-5)$$

$$x + 16 = -5$$

$$x + 16 + (-16) = -5 + (-16)$$

$$x = -21$$

Decimals can be eliminated from equations in a similar manner. To make decimal numbers into whole numbers, the decimal points need to be moved to the right end of the number. This can be done by multiplying each term by the same power of 10.

Example:

$$0.03x + 1.2 = 4 + 0.6x$$

0.03 would need to be multiplied by 100 to become a whole number.

$$100(0.03x) + 100(1.2) = 100(4) + 100(0.6x)$$

$$3x + 120 = 400 + 60x$$

Therefore, all terms must be multiplied by 100.

$$3x + (-3x) + 120 = 400 + 60x + (-3x)$$

$$120 = 400 + 57x$$

$$120 + (-400) = 400 + (-400) + 57x$$

$$-280 = 57x$$

Note: The answer can be left as a fraction.

$$-\frac{280}{57} = x$$

Word problems need to be translated into an equation and then solved:

— Let x represent the quantity being asked for in the problem.

— Write an equation, using x, that describes the situation in the problem.

— Solve the equation.

— Check the solution.

 Test Yourself

Evaluate each of the following when $a = 5$, $b = 4$, $c = -2$, $d = -3$.

1. $3a - 4c$

2. $2a + 3b$

3. $a^2 - 2d^2$

4. $2(a + c)$

5. $d(2a - bc)$

6. $\dfrac{3cd}{3b - cd}$

7. $\dfrac{5a + d}{11c}$

Evaluate each of the following when $a = 0.8$, $b = -3.6$, and $c = 2.2$.

8. $(a + b)c$

9. $ac - b$

10. $b - a$

11. $c - a - b$

12. $a^2 + c$

13. $\dfrac{c}{a} + b$

Evaluate each of the following when $a = \dfrac{-4}{5}$

$b = \dfrac{7}{15}$ $c = \dfrac{-5}{9}$

14. $a + b + c$

15. $ac - b$

16. $c - a - b$

17. $\dfrac{a - c}{b + c}$

18. $a + b - c$

19. a^2

Solve each of the following equations.

20. $x + 8 = -7$

21. $x - 10 = -3$

22. $-2x = -18$

23. $\dfrac{1}{7}x = -9$

24. $4x - 12 = -36$

25. $7x - 15 = -8$

26. $-5x + 6 = 16$

27. $45 - x = 13$

28. $16.3 - 7.2x = -8.18$

29. $3(3x - 1) = -21$

30. $24 - 14x = -8x - 6$

31. $17x + 18 + 7x = 4x - 22$

32. $7x - 6 = -17 + 5x$

33. $2x - 4 = -16 - x$

34. $4(3x + 4) - 20 = 3 + 5x$

35. $x + 66 = -11(x - 16) + x$

36. $5(2x - 4) + 3 = 4x - 5$

37. $3x - 15 = 15 - 3x$

38. $4.8x - 0.83 = -9.95$

39. $3x + 7 - 2x = 4x - 2$

40. $3(x + 5) = 2x - 1$

41. $8x - 10 = -4(x + 3) + 2$

42. $3x + 2.2 - 0.5x = 2.9625$

43. $-(x - 4) = 19x + 8$

44. $6 - 2(x + 3) = 4(x - 1)$

45. $\dfrac{3x}{4} + \dfrac{5}{3} - \dfrac{x}{2} = \dfrac{5}{6}$

46. $-2(2x - 5) + 3(x + 6) = -4(x - 5) - 1$

47. $5x - (2x - 10) = 25$

48. $\dfrac{1}{3}x - \dfrac{4}{5} = \dfrac{8}{15}$

49. $\dfrac{7}{18}x - \dfrac{5}{27} = -\dfrac{8}{9} - \dfrac{1x}{9}$

Write symbols for each of the following phrases.

50. A number increased by 10

51. Subtract 7 from a number

52. The product of a number and negative eight

53. Three less than twice a number

54. The difference of a number and negative three

55. A number divided by 4

56. The sum of twice a number and 11

57. The difference of a number and 1, divided by negative two

58. Three more than a number

59. Four times the sum of a number and 6

60. The sum of eight and a number, decreased by nine

Translate each into an equation and solve:

61. The sum of two times a number x and 7 is 35. What is the value of x?

62. Four times a number is decreased by 20, and the result is 32. What is the number?

63. Five times a number decreased by 40 is 8 more than the number. What is the number?

64. Thirty less than a number is equal to the product of 3 and the sum of the number and 6. What is the number?

65. The product of 3 and the sum of twice a number and 8 is 48. What is the number?

66. If 9 is subtracted from a number divided by 4, the result is -3. What is the number?

67. Three times the sum of a number and 1 is negative 18. What is the number?

68. A number is divided by 3. If the quotient is increased by $\frac{2}{5}$, the result is $1\frac{11}{15}$. What is the number?

69. The sum of 0.75 times a number and 0.95 is 5.2475. What is the number?

70. If the sum of twice a number and triple the same number is decreased by 3, the result is 12. What is the number?

 # Check Yourself

1. $3(5) - 4(-2)$

   $= 3(5) + (-4)(-2)$

   $= 15 + 8$

   $= 23$ **(Evaluating an expression)**

2. $2(5) + 3(4)$

   $= 10 + 12$

   $= 22$ **(Evaluating an expression)**

3. $(5)^2 + (-2)(-3)^2$

   $= 25 + (-2)(9)$

   $= 25 + (-18)$

   $= 7$ **(Evaluating an expression)**

4. $2(5 + (-2))$

   $= 2(3)$

   $= 6$ **(Evaluating an expression)**

For 1 – 19, in general:

— Rewrite subtraction as addition of opposite.

— Use the order of operation rules.

5. $-3\,[\,2\,(5)\,-\,(4)\,(-2)\,]$

    $=-3\,[\,2\,(5)\,+\,(-4)\,(-2)\,]$

    $=-3\,(10+8)$

    $=-3\,(18)$

    $=-54$ **(Evaluating an expression)**

6. $\dfrac{3\,(-2)\,(-3)}{3\,(4)\,-\,(-2)\,(-3)}$ Evaluate the numerator, then the denominator, and then the fraction as a whole.

    $=\dfrac{-6\,(-3)}{12\,+\,(2)\,(-3)}$

    $=\dfrac{18}{12\,+\,(-6)}$

    $=\dfrac{18}{6}$

    $=3$ **(Evaluating an expression)**

7. $\dfrac{5\,(5)\,+\,(-3)}{11\,(-2)}$

    $=\dfrac{25\,+\,(-3)}{-22}$

    $=\dfrac{22}{-22}$

    $=-1$ **(Evaluating an expression)**

8. $[\,0.8\,+\,(-3.6)\,]\,(2.2)$

    $=[\,0.8\,-\,3.6\,]\,(2.2)$

    $=[\,-2.8\,]\,(2.2)$

    $=-6.16$ **(Evaluating an expression)**

9. $(0.8)\,(2.2)\,-\,(-3.6)$

    $=1.76\,-\,(-3.6)$

    $=1.76\,+\,3.6$

    $=5.36$ **(Evaluating an expression)**

10. $-3.6 - 0.8$

    $= -3.6 + (-0.8)$

    $= -4.4$ **(Evaluating an expression)**

11. $2.2 - 0.8 - (-3.6)$

    $= 2.2 + (-0.8) + 3.6$

    $= 1.4 + 3.6$

    $= 5.0$ **(Evaluating an expression)**

12. $(0.8)^2 + (2.2)$

    $= 0.64 + 2.2$

    $= 2.84$ **(Evaluating an expression)**

13. $\dfrac{2.2}{0.8} + (-3.6)$

    $= 2.75 + (-3.6)$

    $= -0.85$ **(Evaluating an expression)**

14. $-\dfrac{4}{5} + \dfrac{7}{15} + \left(-\dfrac{5}{9}\right)$

    $= -\dfrac{36}{45} + \dfrac{21}{45} + \left(-\dfrac{25}{45}\right)$

    $= -\dfrac{61}{45} + \dfrac{21}{45}$

    $= -\dfrac{40}{45}$

    $= -\dfrac{8}{9}$ **(Evaluating an expression)**

15. $\left(-\dfrac{4}{5}\right)\left(-\dfrac{5}{9}\right) - \dfrac{7}{15}$

    $= \dfrac{20}{45} - \dfrac{7}{15}$

    $= \dfrac{20}{45} + \left(-\dfrac{7}{15}\right)$

$$= \frac{20}{45} - \left(\frac{21}{45} \right)$$

$$= -\frac{1}{45} \quad \textbf{(Evaluating an expression)}$$

16. $\quad -\dfrac{5}{9} - \left(-\dfrac{4}{5} \right) - \dfrac{7}{15}$

$$= -\frac{5}{9} + \frac{4}{5} + \left(-\frac{7}{15} \right)$$

$$= -\frac{25}{45} + \frac{36}{45} - \frac{21}{45}$$

$$= -\frac{10}{45}$$

$$= -\frac{2}{9} \quad \textbf{(Evaluating an expression)}$$

17. $\quad \dfrac{-\dfrac{4}{5} - \left(-\dfrac{5}{9} \right)}{\dfrac{7}{15} + \left(-\dfrac{5}{9} \right)}$

$$= \frac{-\dfrac{36}{45} + \dfrac{25}{45}}{\dfrac{21}{45} + \left(-\dfrac{25}{45} \right)}$$

$$= \frac{-\dfrac{11}{45}}{-\dfrac{4}{45}}$$

$$= -\frac{11}{45} \div \left(-\frac{4}{45} \right)$$

$$= -\frac{11}{45} \cdot -\frac{45}{4}$$

$$= -\frac{11}{4}$$

$$= 2\frac{3}{4} \quad \textbf{(Evaluating an expression)}$$

18. $\quad -\dfrac{4}{5} + \dfrac{7}{15} - \left(-\dfrac{5}{9} \right)$

$$= -\frac{36}{45} + \frac{21}{45} + \frac{25}{45}$$

$$= -\frac{36}{45} + \frac{46}{45}$$

$$= \frac{10}{45}$$

$$= \frac{2}{9} \quad \textbf{(Evaluating an expression)}$$

19. $\left(-\frac{4}{5}\right)^2$

$$= -\frac{4}{5} \cdot -\frac{4}{5}$$

$$= \frac{16}{25} \quad \textbf{(Evaluating an expression)}$$

20. $\qquad x + 8 = -7$

$$x + 8 + (-8) = -7 + (-8)$$

$$x = -15 \quad \textbf{(Solving an equation)}$$

21. $\qquad x - 10 = -3$

$$x + (-10) = -3$$

$$x + (-10) + 10 = -3 + 10$$

$$x = 7 \quad \textbf{(Solving an equation)}$$

22. $\quad -2x = -18$

$$\frac{-2x}{-2} = \frac{-18}{-2}$$

$$x = 9 \quad \textbf{(Solving an equation)}$$

23. $\qquad \frac{1}{7}x = -9$

$$(7)\frac{1}{7}x = (-9)(7)$$

$$x = -63 \quad \textbf{(Solving an equation)}$$

24. $\qquad 4x - 12 = -36$

$$4x + (-12) = -36$$

$$4x + (-12) + 12 = -36 + 12$$

$$4x = -24$$

$$x = -6 \quad \textbf{(Solving an equation)}$$

25. $\qquad 7x - 15 = -8$

$$7x + (-15) = -8$$

$$7x + (-15) + 15 = -8 + 15$$

$$7x = 7$$

$$x = 1 \quad \textbf{(Solving an equation)}$$

26. $\qquad -5x + 6 = 16$

$$-5x + 6 + (-6) = 16 + (-6)$$

$$-5x = 10$$

$$x = -2 \quad \textbf{(Solving an equation)}$$

27. $\qquad 45 - x = 13$

$$45 + (-x) = 13$$

$$45 + (-45) + (-x) = 13 + (-45)$$

$$-x = -32$$

$$x = 32 \quad \textbf{(Solving an equation)}$$

28. $\qquad 16.3 + (-7.2x) = -8.18$

$$16.3 + (-16.3) + (-7.2x) = -8.18 + (-16.3)$$

$$-7.2x = -24.48$$

$$x = 3.4 \quad \textbf{(Solving an equation)}$$

29. $\qquad 9x - 3 = -21$

$$9x + (-3) + 3 = -21 + 3$$

$$9x = -18$$

$$x = -2 \quad \textbf{(Solving an equation)}$$

30. $\qquad 24 + (-14x) = -8x + (-6)$

$$24 + (-14x) + 14x = -8x + 14x + (-6)$$

$$24 = 6x + (-6)$$

$$24 + 6 = 6x + (-6) + 6$$

$$30 = 6x$$

$$5 = x \quad \textbf{(Solving an equation)}$$

31.
$$24x + 18 = 4x + (-22)$$

$$24x + (-4x) + 18 = 4x + (-4x) + (-22)$$

$$20x + 18 = -22$$

$$20x + 18 + (-18) = -22 + (-18)$$

$$20x = -40$$

$$x = -2 \quad \textbf{(Solving an equation)}$$

32.
$$7x + (-6) = -17 + 5x$$

$$7x + (-5x) + (-6) = -17 + 5x + (-5x)$$

$$2x + (-6) = -17$$

$$2x = -11$$

$$x = -\frac{11}{2} \quad \textbf{(Solving an equation)}$$

33.
$$2x + (-4) = -16 + (-x)$$

$$2x + x + (-4) = -16 + (-x) + x$$

$$3x + (-4) = -16$$

$$3x + (-4) + 4 = -16 + 4$$

$$3x = -12$$

$$x = -4 \quad \textbf{(Solving an equation)}$$

34.
$$12x + 16 + (-20) = 3 + 5x$$

$$12x + (-4) = 3 + 5x$$

$$12x + (-5x) + (-4) = 3 + 5x + (-5x)$$

$$7x + (-4) = 3$$

$$7x + (-4) + 4 = 3 + 4$$

$$7x = 7$$

$$x = 1 \quad \textbf{(Solving an equation)}$$

35.
$$x + 66 = -11x + 176 + x$$
$$x + 66 = -10x + 176$$
$$x + 10x + 66 = -10x + 10x + 176$$
$$11x + 66 = 176$$
$$11x + 66 + (-66) = 176 + (-66)$$
$$11x = 110$$
$$x = 10 \quad \textbf{(Solving an equation)}$$

36.
$$5(2x + (-4)) + 3 = 4x + (-5)$$
$$10x + (-20) + 3 = 4x + (-5)$$
$$10x + (-17) = 4x + (-5)$$
$$10x + (-4x) + (-17) = 4x + (-4x) + (-5)$$
$$6x + (-17) = -5$$
$$6x + (-17) + 17 = -5 + 17$$
$$6x = 12$$
$$x = 2 \quad \textbf{(Solving an equation)}$$

37.
$$3x + (-15) = 15 + (-3x)$$
$$3x + 3x + (-15) = 15 + (-3x) + 3x$$
$$6x + (-15) = 15$$
$$6x + (-15) + 15 = 15 + 15$$
$$6x = 30$$
$$x = 5 \quad \textbf{(Solving an equation)}$$

38.
$$4.8x + (-0.83) = -9.95 \qquad \text{Multiply every term by 100.}$$
$$480x + (-83) = -995$$
$$480x + (-83) + 83 = -995 + 83$$
$$480x = -912$$
$$x = -1.9 \quad \textbf{(Solving an equation)}$$

39.
$$3x + 7 + (-2x) = 4x + (-2)$$
$$x + 7 = 4x + (-2)$$

$$x + (-x) + 7 = 4x + (-x) + (-2)$$

$$7 = 3x + (-2)$$

$$7 + 2 = 3x + (-2) + 2$$

$$9 = 3x$$

$$3 = x \quad \textbf{(Solving an equation)}$$

40.
$$3(x + 5) = 2x + (-1)$$

$$3x + 15 = 2x + (-1)$$

$$3x + (-2x) + 15 = 2x + (-2x) + (-1)$$

$$x + 15 = -1$$

$$x + 15 + (-15) = -1 + (-15)$$

$$x = -16 \quad \textbf{(Solving an equation)}$$

41.
$$8x + (-10) = -4(x + 3) + 2$$

$$8x + (-10) = -4x + (-12) + 2$$

$$8x + (-10) = -4x + (-10)$$

$$8x + 4x + (-10) = -4x + 4x + (-10)$$

$$12x + (-10) = -10$$

$$12x + (-10) + 10 = -10 + 10$$

$$12x = 0$$

$$x = 0 \quad \textbf{(Solving an equation)}$$

42.
$$3x + 2.2 - 0.5x = 2.9625 \quad \text{Multiply every term by 10,000.}$$

$$30,000x + 22,000 - 5,000x = 29,625$$

$$25,000x + 22,000 = 29,625$$

$$25,000x + 22,000 + (-22,000) = 29,625 + (-22,000)$$

$$25,000x = 7625$$

$$x = 0.305 \quad \textbf{(Solving an equation)}$$

43.
$$-(x + (-4)) = 19x + 8$$

$$-x + 4 = 19x + 8$$

$$-x + x + 4 = 19x + x + 8$$

$$4 = 20x + 8$$

$$4 + (-8) = 20x + 8 + (-8)$$

$$-4 = 20x$$

$$\frac{-4}{20} = x$$

$$-\frac{1}{5} = x \quad \textbf{(Solving an equation)}$$

44. $\quad 6 - 2(x + 3) = 4(x - 1)$

$$6 + (-2)(x + 3) = 4(x + (-1))$$

$$6 + (-2x) + (-6) = 4x + (-4)$$

$$-2x = 4x + (-4)$$

$$-2x + 2x = 4x + 2x + (-4)$$

$$0 = 6x + (-4)$$

$$4 = 6x$$

$$\frac{4}{6} = x$$

$$x = \frac{2}{3} \quad \textbf{(Solving an equation)}$$

45. $\quad \dfrac{3x}{4} + \dfrac{5}{3} - \dfrac{x}{2} = \dfrac{5}{6}$

$$9x + 20 - 6x = 10 \qquad \text{Multiply every term by 12.}$$

$$3x + 20 = 10$$

$$3x + 20 + (-20) = 10 + (-20)$$

$$3x = -10$$

$$x = -\frac{10}{3} \quad \textbf{(Solving an equation)}$$

46. $\quad -2(2x - 5) + 3(x + 6) = -4(x - 5) - 1$

$$-2(2x + (-5)) + 3(x + 6) = -4(x + (-5)) + (-1)$$

$$-4x + 10 + 3x + 18 = -4x + 20 + (-1)$$

$$-x + 28 = -4x + 19$$

$$-x + 4x + 28 = -4x + 4x + 19$$

$$3x + 28 = 19$$

$$3x + 28 + (-28) = 19 + (-28)$$

$$3x = -9$$

$$x = -3 \quad \textbf{(Solving an equation)}$$

47. $\quad 5x - (2x - 10) = 25$

$$5x - (2x + (-10)) = 25$$

$$5x + (-2x) + 10 = 25$$

$$3x + 10 + (-10) = 25 + (-10)$$

$$3x = 15$$

$$x = 5 \quad \textbf{(Solving an equation)}$$

48. $\quad \dfrac{1}{3}x - \dfrac{4}{5} = \dfrac{8}{15} \qquad$ Multiply every term by 15.

$$5x - 12 = 8$$

$$5x + (-12) = 8$$

$$5x + (-12) + 12 = 8 + 12$$

$$5x = 20$$

$$x = 4 \quad \textbf{(Solving an equation)}$$

49. $\quad \dfrac{7}{18}x - \dfrac{5}{27} = -\dfrac{8}{9} + \left(-\dfrac{1}{9}x\right) \qquad$ Multiply every term by 54.

$$21x + (-10) = -48 + (-6x)$$

$$21x + 6x + (-10) = -48 + (-6x) + (6x)$$

$$27x + (-10) = -48$$

$$27x + (-10) + (10) = -48 + 10$$

$$27x = -38$$

$$x = -\dfrac{38}{27} \quad \textbf{(Solving an equation)}$$

50. $\quad x + 10$ **(Translation into an algebraic expression)**

51. $\quad x - 7$ **(Translation into an algebraic expression)**

52. $\quad -8x$ **(Translation into an algebraic expression)**

53. $2x - 3$ **(Translation into an algebraic expression)**

54. $x - (-3)$ **(Translation into an algebraic expression)**

55. $\dfrac{x}{4}$ **(Translation into an algebraic expression)**

56. $2x + 11$ **(Translation into an algebraic expression)**

57. $\dfrac{x - 1}{-2}$ **(Translation into an algebraic expression)**

58. $x + 3$ **(Translation into an algebraic expression)**

59. $4(x + 6)$ **(Translation into an algebraic expression)**

60. $(8 + x) - 9$ **(Translation into an algebraic expression)**

61.
$$2x + 7 = 35$$
$$2x + 7 + (-7) = 35 + (-7)$$
$$2x = 28$$
$$x = 14 \quad \textbf{(Solving word problems)}$$

62.
$$4x - 20 = 32$$
$$4x + (-20) = 32$$
$$4x + (-20) + (20) = 32 + 20$$
$$4x = 52$$
$$x = 13 \quad \textbf{(Solving word problems)}$$

63.
$$5x - 40 = x + 8$$
$$5x + (-40) = x + 8$$
$$5x + (-40) + 40 = x + 8 + 40$$
$$5x = x + 48$$
$$5x + (-x) = x + (-x) + 48$$
$$4x = 48$$
$$x = 12 \quad \textbf{(Solving word problems)}$$

64.
$$x - 30 = 3(x + 6)$$
$$x + (-30) = 3x + 18$$
$$x + (-x) + (-30) = 3x + (-x) + 18$$

$$-30 = 2x + 18$$

$$-30 - 18 = 2x + 18 + (-18)$$

$$-48 = 2x$$

$$-24 = x \quad \textbf{(Solving word problems)}$$

65.
$$3(2x + 8) = 48$$

$$6x + 24 = 48$$

$$6x + 24 + (-24) = 48 + (-24)$$

$$6x = 24$$

$$x = 4 \quad \textbf{(Solving word problems)}$$

66.
$$\frac{x}{4} - 9 = -3$$

$$\frac{x}{4} + (-9) = -3$$

$$\frac{x}{4} + (-9) + 9 = -3 + 9$$

$$\frac{x}{4} = 6$$

$$x = 24 \quad \textbf{(Solving word problems)}$$

67.
$$3(x + 1) = -18$$

$$3x + 3 = -18$$

$$3x + 3 + (-3) = -18 + (-3)$$

$$3x = -21$$

$$x = -7 \quad \textbf{(Solving word problems)}$$

68.
$$\frac{x}{3} + \frac{2}{5} = 1\frac{11}{15}$$

$$\frac{x}{3} + \frac{2}{5} = \frac{26}{15}$$

This could also be done by multiplying every term by 15 to eliminate the fractions.

$$\frac{x}{3} + \frac{6}{15} = \frac{26}{15}$$

$$\frac{x}{3} + \frac{6}{15} + \left(-\frac{6}{15}\right) = \frac{26}{15} + \left(-\frac{6}{15}\right)$$

$$\frac{x}{3} = \frac{20}{15}$$

$x = 4$ **(Solving word problems)**

69. $0.75x + 0.95 = 5.2475$

$0.75x + 0.95 + (-0.95) = 5.2475 + (-0.95)$

$0.75x = 4.2975$ This could also be done by multiplying every term by 10,000 to eliminate the decimals. In this case the final answer would be the fraction $5\frac{73}{100}$.

$x = 5.73$ **(Solving word problems)**

70. $(2x + 3x) - 3 = 12$

$5x + (-3) = 12$

$5x + (-3) + 3 = 12 + 3$

$5x = 15$

$x = 3$ **(Solving word problems)**

Grade Yourself

Circle the question numbers that you had incorrect. Then indicate the number of questions you missed. If you answered more than three questions incorrectly, you need to focus on that topic. (If a topic has less than three questions and you had at least one wrong, we suggest you study that topic also. Read your textbook or a review book, or ask your teacher for help.)

Subject: *Introduction to Polynomials*

Topic	Question Numbers	Number Incorrect
Evaluating an expression	1, 2, 3, 4, 5, 6, 7, 8, 9, 10, 11, 12, 13, 14, 15, 16, 17, 18, 19	
Solving an equation	20, 21, 22, 23, 24, 25, 26, 27, 28, 29, 30, 31, 32, 33, 34, 35, 36, 37, 38, 39, 40, 41, 42, 43, 44, 45, 46, 47, 48, 49	
Translation into an algebraic expression	50, 51, 52, 53, 54, 55, 56, 57, 58, 59, 60	
Solving word problems	61, 62, 63, 64, 65, 66, 67, 68, 69, 70	

Polynomials

9

Brief Yourself

This chapter reviews the rules used to add, subtract, multiply, and divide polynomials. Division will be restricted to division by a monomial only.

Chapter 8 contained the definitions of the following words: constant, variable, term, coefficient, and expression. A quick review of these definitions might be helpful.

Polynomial is another name for an algebraic expression. It is the sum or difference of terms.

    Examples: $3x + 4$ $x^2 + 7xy + 6$

Polynomials with one term, two terms or three terms have the following special names:

    monomial — a polynomial with one term: x^3

    binomial — a polynomial with two terms: $6y^2 - 7$

    trinomial — a polynomial with three terms: $x^3 - y + 11$

Addition and Subtraction of Polynomials

    — Combine like terms by adding or subtracting the numerical coefficients. The variable parts should remain the same.
    — All rules for fractions, decimals and signed numbers must be applied.

    Example:

$(4x^2 + 6x + 7) + (7x^2 - 5x - 11)$	Note: The parentheses are unnecessary.
$(4x^2 + 6x + 7) + (7x^2 + (-5x) + (-11))$	Rewrite subtraction as addition of the opposite.
$4x^2 + 7x^2 + 6x + (-5x) + 7 + (-11)$	Use commutative property to change the order — like terms are next to each other.
$11x^2 + x - 4$	It is unnecessary to write down the commutative step.

When there are no like terms, the problem is completely simplified.

    Example:

$(4x + 5) - (6x - 9)$	When subtracting, be careful of the subtraction signs.

$(4x + 5) - (6x + (-9))$ The minus sign in between the parentheses changes the sign of all of the terms in the second polynomial.

$4x + 5 + (-6x) + 9$

$4x + (-6x) + 5 + 9$

$-2x + 14$

NOTE: At this point, many students find it unnecessary to rewrite subtraction as addition of the opposite. Personal preference should guide you. The previous example would look like the following if the subtraction were not rewritten:

$(4x + 5) - (6x - 9)$

$4x + 5 - 6x + 9$

$-2x + 14$

Multiplication

(Monomial)(Monomial)
— Use commutative property (not necessary to actually write this step).
— Multiply the coefficients.
— Use the exponent rules for the variables with like bases.
— Use appropriate sign rules.

Example:

$(3xy)(-4x^2y) = (3)(-4)(x)(x^2)(y)(y)$

$= -12x^3y^2$

(Monomial)(Polynomial)
— Rewrite any subtractions within the polynomial.
— Use distributive property to multiply each term of the polynomial by the monomial.

Example:

$-3x(4x^2 - 5x - 3) = -3x(4x^2 + -5x + -3)$

$= (-3x)(4x^2) + (-3x)(-5x) + (-3x)(-3)$

$= -12x^3 + 15x^2 + 9x$

(Polynomial)(Polynomial)
— If a polynomial is raised to a power, rewrite in expanded form.
— Rewrite any subtraction symbols as addition of the opposite.
— Use distributive property to multiply each term of the second polynomial by each term of the first polynomial.

Example:

$(x+4)(2x+8)$

$x(2x) + x(8) + 4(2x) + 4(8)$ First distribute the x, then the 4.

$2x^2 + 8x + 8x + 32$ Multiply the coefficients, add the exponents of the like bases.

Example:

$2x^2 + 16x + 32$ Combine like terms.

$(x^2 + 2x - 3)^2$

$(x^2 + 2x - 3)(x^2 + 2x - 3)$ Rewrite in expanded form.

$(x^2 + 2x + -3)(x^2 + 2x + -3)$ Rewrite subtraction.

First distribute the x^2, then the $2x$ and then the -3.

$x^2(x^2) + x^2(2x) + x^2(-3) + 2x(x^2) + 2x(2x) + 2x(-3) + (-3)(x^2) + (-3)(2x) + (-3)(-3)$

$x^4 + 2x^3 + -3x^2 + 2x^3 + 4x^2 + -6x + -3x^2 + -6x + 9$ Combine like terms.

$x^4 + 4x^3 - 2x^2 - 12x + 9$

Division

Polynomial ÷ Monomial   — Rewrite any subtractions within the polynomial.
                                 — Write each term of the numerator with its own denominator.
                                 — Simplify each term using exponent rules when necessary.

Example:

$\dfrac{16x^4y^3 + 8x^3y^4 - 4x^2y^2}{4x^2y^2}$

$\dfrac{16x^4y^3}{4x^2y^2} + \dfrac{8x^3y^4}{4x^2y^2} + \dfrac{-4x^2y^2}{4x^2y^2}$ Write each term of the numerator with its own denominator.

$4x^2y + 2xy^2 - 1$ Divide coefficients, subtract exponents of like bases.

Test Yourself

Add each of the following.

1. $(3x^2 + 2x + 9) + (4x^2 + 5x + 3)$

2. $(2x + 3) + (4x - 2)$

3. $(-2x^2 - 4x - 8) + (6x + 3)$

4. $(9x^2 + 3x - 11) + (3x^2 - 12x - 5)$

5. $(5x + 8) + (-6x - 13)$

6. $(4x^2 + 9x + 1) + (2x^2 + 6x - 2)$

7. $(-2x + 6) + (-4x^2 - 7x - 9)$

8. $(7x^2 - 6) + (-2x^2 - 5x)$

9. $(4x^2 - 2x + 6) + (-7x^2 - 3x - 9)$

10. $(6x^3 + x - 6) + (-2x^3 - 3x^2)$

11. $(3x^2y - 4xy + y) + (x^2y + 2xy + 3y)$

12. $(6x^3 - 4x^2 + x - 9) + (-x^3 - 3x^2 - x + 7)$

13. $(x^2y + 8x^2 - 2xy^2) + (-x^2y - 11x^2 + 5xy^2)$

14. $(-7x^3 - 4x^2 + 6) + (5x^3 + 3x - 8)$

15. $(x^2 + xy - y^2) + (2x^2 - 3xy + y^2)$

Subtract each of the following.

16. $(3x + 4) - (2x + 2)$

17. $(6x + 3) - (4x - 2)$

18. $(-2x - 3) - (5x + 7)$

19. $(-x + 4) - (-x - 9)$

20. $(9x^2 + 3x - 5) - (3x^2 + 5)$

21. $(-2x^2 + 4x - 9) - (5x^2 - 3x + 7)$

22. $(x^2 - 6x) - (2x^2 + 4)$

23. $(9x - 6) - (-2x^2 - 5x - 3)$

24. $(x^2y + 6x - 3) - (-4x^2y - 5x - 1)$

25. $(-x^3 - 3xy + y^2) - (4xy - 7y^2)$

26. Subtract $-9x - 4$ from $-5x + 6$

27. Subtract $x^2 + 6x - 9$ from $-5x^2 - 6x + 2$

28. Subtract $-2x - 8$ from $4x^2 - 6x + 5$

29. Subtract $x^2 + 2x - 9$ from $-5x^2 - 3$

30. Subtract $5x^3 - 4x^2 - 6x - 1$ from $2x^3 - 6x^2 - x + 1$

Multiply each of the following.

31. $2x(3x^2 + 4)$

32. $-5x(x + 2)$

33. $3x^2(4x^2 + 6x - 2)$

34. $3(x - 5)$

35. $-4x(-2x + 3)$

36. $(3x - 8)(2x + 5)$

37. $(4x - 2y)(2y - 5x)$

38. $(2x + 1)(x + 5)$

39. $(4 - x)(2x + 3)$

40. $(2x + 3)^2$

41. $(3x + 4)(3x - 4)$

42. $(2x - 5)(3x^2 - 4x + 7)$

43. $(4x^2 - 6)(3x + 1)$

44. $(4x^2 + 9x - 2)(x - 3)$

45. $(x^2 - 2x + 3)(x^2 - 4)$

Divide each of the following.

46. $\dfrac{2x + 4}{2}$

47. $\dfrac{4x - 6}{-2}$

48. $\dfrac{12x^2 - 6x + 3}{3}$

49. $\dfrac{-12x^3 + 6x^2 - 15x}{-3x}$

50. $\dfrac{3xy + 9x^3y^2}{3xy}$

51. $\dfrac{45xyz + 30xy - 60xz}{15x}$

52. $\dfrac{72xyz - 48xy - 64xz}{-8x}$

53. $\dfrac{-16xy + 12xz - 24xw}{4x}$

54. $\dfrac{32x^2y + 48x^3y^2 - 56xy^3}{-6y}$

55. $\dfrac{75x^2yz - 15xy^2z - 90xyz^2}{15xyz}$

Perform the indicated operations.

56. $x\,[\,3 - 2\,(x - 1)\,(x + 1)\,]$

57. $3x\,(x + 2)\,(x - 3)$

58. $3x\,[\,(x + 2) + (x - 3)\,]$

59. $(2x^2 - 5x) - [\,(3x - 2x^2) - (5x^2 - 2x)\,]$

60. $(x^2 - 2x + 1) - [\,(2x^2 - 3) + (-3x^2 + 7)\,]$

61. $(4x + 1)\,(3x - 2) - (3x + 1)\,(4x - 2)$

62. $6x\,(x^2 - 5) - 6x\,(x^2 + 5x + 5)$

63. $x - [\,3x^2y - (x - 1)\,] - [\,-y\,(2x^2 + y)\,]$

64. $(x^2 + 4) - [\,(x^2 - 5) - (3x^2 + 1)\,]$

65. $\dfrac{9x^2y + 3x}{3x}$

66. $5x\,(3x - 2) - (2x - 3)^2$

Check Yourself

1. $(3x^2 + 2x + 9) + (4x^2 + 5x + 3)$

   $3x^2 + 4x^2 + 2x + 5x + 9 + 3$ Commutative Property — not necessary to show.

   $7x^2 + 7x + 12$ Collect like terms. (**Addition of polynomials**)

2. $(2x + 3) + (4x - 2)$

   $(2x + 3) + (4x + -2)$ Rewrite subtraction.

   $2x + 4x + 3 + -2$ Commutative Property.

   $6x + 1$ Collect like terms. (**Addition of polynomials**)

3. $(-2x^2 - 4x - 8) + (6x + 3)$

   $(-2x^2 + -4x + -8) + (6x + 3)$

   $-2x^2 + -4x + 6x + -8 + 3$

   $-2x^2 + 2x - 5$ (**Addition of polynomials**)

4. $(9x^2 + 3x - 11) + (3x^2 - 12x - 5)$

   $(9x^2 + 3x + -11) + (3x^2 + -12x + -5)$

   $9x^2 + 3x^2 + 3x + -12x + -11 + -5$

$12x^2 + -9x + -16$

$12x^2 - 9x - 16$ (**Addition of polynomials**)

5. $(5x + 8) + (-6x - 13)$

$(5x + 8) + (-6x + -13)$

$5x + -6x + 8 + -13$

$-x + -5$

$-x - 5$ (**Addition of polynomials**)

6. $(4x^2 + 9x + 1) + (2x^2 + 6x - 2)$

$(4x^2 + 9x + 1) + (2x^2 + 6x + -2)$

$4x^2 + 2x^2 + 9x + 6x + 1 + -2$

$6x^2 + 15x + -1$

$6x^2 + 15x - 1$ (**Addition of polynomials**)

7. $(-2x + 6) + (-4x^2 - 7x - 9)$

$(-2x + 6) + (-4x^2 + -7x + -9)$

$-4x^2 + -2x + -7x + 6 + -9$

$-4x^2 + -9x + -3$

$-4x^2 - 9x - 3$ (**Addition of polynomials**)

8. $(7x^2 - 6) + (-2x^2 - 5x)$

$7x^2 - 6 + -2x^2 - 5x$

$5x^2 - 5x - 6$ (**Addition of polynomials**)

9. $(4x^2 - 2x + 6) + (-7x^2 - 3x - 9)$

$4x^2 - 2x + 6 + -7x^2 - 3x - 9$

$-3x^2 - 5x - 3$ (**Addition of polynomials**)

10. $(6x^3 + x - 6) + (-2x^3 - 3x^2)$

$6x^3 + x - 6 + -2x^3 - 3x^2$

$4x^3 - 3x^2 + x - 6$ (**Addition of polynomials**)

11. $(3x^2y - 4xy + y) + (x^2y + 2xy + 3y)$

    $3x^2y - 4xy + y + x^2y + 2xy + 3y$

    $4x^2y - 2xy + 4y$ (**Addition of polynomials**)

12. $(6x^3 - 4x^2 + x - 9) + (-x^3 - 3x^2 - x + 7)$

    $6x^3 - 4x^2 + x - 9 + -x^3 - 3x^2 - x + 7$

    $5x^3 - 7x^2 - 2$ (**Addition of polynomials**)

13. $(x^2y + 8x^2 - 2xy^2) + (-x^2y - 11x^2 + 5xy^2)$

    $x^2y + 8x^2 - 2xy^2 + -x^2y - 11x^2 + 5xy^2$

    $-3x^2 + 3xy^2$ (**Addition of polynomials**)

14. $(-7x^3 - 4x^2 + 6) + (5x^3 + 3x - 8)$

    $-7x^3 - 4x^2 + 6 + 5x^3 + 3x - 8$

    $-2x^3 - 4x^2 + 3x - 2$

15. $(x^2 + xy - y^2) + (2x^2 - 3xy + y^2)$

    $x^2 + xy - y^2 + 2x^2 - 3xy + y^2$

    $3x^2 - 2xy$ (**Addition of polynomials**)

16. $(3x + 4) - (2x + 2)$

    $(3x + 4) + -1(2x + 2)$ Rewrite subtraction.

    $3x + 4 + -2x + -2$ Distribute -1.

    $x + 2$ Collect like terms. (**Subtraction of polynomials**)

17. $(6x + 3) - (4x - 2)$

    $(6x + 3) - (4x + -2)$ Rewrite subtraction inside parentheses.

    $(6x + 3) + -1(4x + -2)$ Rewrite subtraction between parentheses.

    $6x + 3 + -4x + 2$ Distribute -1.

    $2x + 5$ Collect like terms. (**Subtraction of polynomials**)

18. $(-2x - 3) - (5x + 7)$

    $(-2x + -3) - (5x + 7)$ Rewrite subtraction inside parentheses.

    $(-2x + -3) + -1(5x + 7)$ Rewrite subtraction between parentheses.

$-2x + -3 + -5x + -7$ Distribute -1.

$-7x + -10$

$-7x - 10$ **(Subtraction of polynomials)**

19. $(-x + 4) - (-x - 9)$

$-x + 4 + x + 9$ Distribute.

13   **(Subtraction of polynomials)**

20. $(9x^2 + 3x - 5) - (3x^2 + 5)$

$9x^2 + 3x - 5 - 3x^2 - 5$

$6x^2 + 3x - 10$ **(Subtraction of polynomials)**

21. $(-2x^2 + 4x - 9) - (5x^2 - 3x + 7)$

$-2x^2 + 4x - 9 - 5x^2 + 3x - 7$

$-7x^2 + 7x - 16$ **(Subtraction of polynomials)**

22. $(x^2 - 6x) - (2x^2 + 4)$

$x^2 - 6x - 2x^2 - 4$

$-x^2 - 6x - 4$ **(Subtraction of polynomials)**

23. $(9x - 6) - (-2x^2 - 5x - 3)$

$9x - 6 + 2x^2 + 5x + 3$

$2x^2 + 14x - 3$ **(Subtraction of polynomials)**

24. $(x^2y + 6x - 3) - (-4x^2y - 5x - 1)$

$x^2y + 6x - 3 + 4x^2y + 5x + 1$

$5x^2y + 11x - 2$ **(Subtraction of polynomials)**

25. $(-x^3 - 3xy + y^2) - (4xy - 7y^2)$

$-x^3 - 3xy + y^2 - 4xy + 7y^2$

$-x^3 + 8y^2 - 7xy$ **(Subtraction of polynomials)**

26.   Subtract $-9x - 4$ from $-5x + 6$

$(-5x + 6) - (-9x - 4)$ Write with parentheses.

$-5x + 6 + 9x + 4$ Distribute.

$4x + 10$ **(Subtraction of polynomials)**

27.   Subtract $x^2 + 6x - 9$ from $-5x^2 - 6x + 2$

$(-5x^2 - 6x + 2) - (x^2 + 6x - 9)$

$-6x^2 - 12x + 11$ **(Subtraction of polynomials)**

28.   Subtract $-2x - 8$ from $4x^2 - 6x + 5$

$(4x^2 - 6x + 5) - (-2x - 8)$

$4x^2 - 6x + 5 + 2x + 8$

$4x^2 - 4x + 13$ **(Subtraction of polynomials)**

29.   Subtract $x^2 + 2x - 9$ from $-5x^2 - 3$

$(-5x^2 - 3) - (x^2 + 2x - 9)$

$-5x^2 - 3 - x^2 - 2x + 9$

$-6x^2 - 2x + 6$ **(Subtraction of polynomials)**

30.   Subtract $5x^3 - 4x^2 - 6x - 1$ from $2x^3 - 6x^2 - x + 1$

$(2x^3 - 6x^2 - x + 1) - (5x^3 - 4x^2 - 6x - 1)$

$2x^3 - 6x^2 - x + 1 - 5x^3 + 4x^2 + 6x + 1$

$-3x^3 - 2x^2 + 5x + 2$ **(Subtraction of polynomials)**

31. $2x(3x^2 + 4)$

$2x(3x^2) + 2x(4)$ Distribute.

$6x^3 + 8x$ **(Multiplication of polynomials)**

32. $-5x(x + 2)$

$-5x(x) + -5x(2)$ Distribute.

$-5x^2 - 10x$ **(Multiplication of polynomials)**

33. $3x^2(4x^2 + 6x - 2)$

$3x^2(4x^2) + 3x^2(6x) + 3x^2(-2)$

$12x^4 + 18x^3 - 6x^2$ **(Multiplication of polynomials)**

34. $3(x-5)$

$3(x) - 3(5)$

$3x - 15$ **(Multiplication of polynomials)**

35. $-4x(-2x+3)$

$-4x(-2x) + -4x(3)$

$8x^2 + -12x$

$8x^2 - 12x$ **(Multiplication of polynomials)**

36. $(3x-8)(2x+5)$

$3x(2x) + 3x(5) - 8(2x) - 8(5)$ Distribute first $3x$, then -8.

$6x^2 + 15x - 16x - 40$

$6x^2 - x - 40$ Collect like terms. **(Multiplication of polynomials)**

37. $(4x-2y)(2y-5x)$

$4x(2y) + 4x(-5x) - 2y(2y) - 2y(-5x)$

$8xy - 20x^2 - 4y^2 + 10xy$

$-20x^2 + 18xy - 4y^2$ **(Multiplication of polynomials)**

38. $(2x+1)(x+5)$

$2x(x) + 2x(5) + 1(x) + 1(5)$

$2x^2 + 10x + x + 5$

$2x^2 + 11x + 5$ **(Multiplication of polynomials)**

39. $(4-x)(2x+3)$

$4(2x) + 4(3) - x(2x) - x(3)$

$8x + 12 - 2x^2 - 3x$

$-2x^2 + 5x + 12$ **(Multiplication of polynomials)**

40. $(2x+3)^2$

$(2x+3)(2x+3)$

$2x(2x) + 2x(3) + 3(2x) + 3(3)$

$4x^2 + 6x + 6x + 9$

$4x^2 + 12x + 9$ **(Multiplication of polynomials)**

41. $(3x + 4)\,(3x - 4)$

$3x\,(3x) + 3x\,(-4) + 4\,(3x) + 4\,(-4)$

$9x^2 - 12x + 12x - 16$

$9x^2 - 16$ **(Multiplication of polynomials)**

42. $(2x - 5)\,(3x^2 - 4x + 7)$

$2x\,(3x^2) + 2x\,(-4x) + 2x\,(7) - 5\,(3x^2) - 5\,(-4x) - 5\,(7)$

$6x^3 - 8x^2 + 14x - 15x^2 + 20x - 35$

$6x^3 - 23x^2 + 34x - 35$ **(Multiplication of polynomials)**

43. $(4x^2 - 6)\,(3x + 1)$

$4x^2\,(3x) + 4x^2\,(1) - 6\,(3x) - 6\,(1)$

$12x^3 + 4x^2 - 18x - 6$ **(Multiplication of polynomials)**

44. $(4x^2 + 9x - 2)\,(x - 3)$

$4x^2\,(x) + 4x^2\,(-3) + 9x\,(x) + 9x\,(-3) - 2\,(x) - 2\,(-3)$

$4x^3 - 12x^2 + 9x^2 - 27x - 2x + 6$

$4x^3 - 3x^2 - 29x + 6$ **(Multiplication of polynomials)**

45. $(x^2 - 2x + 3)\,(x^2 - 4)$

$x^2\,(x^2) + x^2\,(-4) - 2x\,(x^2) - 2x\,(-4) + 3\,(x^2) + 3\,(-4)$

$x^4 - 4x^2 - 2x^3 + 8x + 3x^2 - 12$

$x^4 - 2x^3 - x^2 + 8x - 12$ **(Multiplication of polynomials)**

46. $\dfrac{2x + 4}{2}$

$\dfrac{2x}{2} + \dfrac{4}{2}$ Write each term of the numerator with its own denominator.

$x + 2$ **(Division of polynomials)**

47. $\dfrac{4x - 6}{-2}$

$\dfrac{4x}{-2} - \dfrac{6}{-2}$

$-2x + 3$ **(Division of polynomials)**

48. $\dfrac{12x^2 - 6x + 3}{3}$

$\dfrac{12x^2}{3} - \dfrac{6x}{3} + \dfrac{3}{3}$

$4x^2 - 2x + 1$ **(Division of polynomials)**

49. $\dfrac{-12x^3 + 6x^2 - 15x}{-3x}$

$\dfrac{-12x^3}{-3x} + \dfrac{6x^2}{-3x} - \dfrac{15x}{-3x}$ Use the exponent rules when dividing like bases.

$4x^2 - 2x + 5$ **(Division of polynomials)**

50. $\dfrac{3xy + 9x^3y^2}{3xy}$

$\dfrac{3xy}{3xy} + \dfrac{9x^3y^2}{3xy}$

$1 + 3x^2y$ **(Division of polynomials)**

51. $\dfrac{45xyz + 30xy - 60xz}{15x}$

$\dfrac{45xyz}{15x} + \dfrac{30xy}{15x} - \dfrac{60xz}{15x}$

$3yz + 2y - 4z$ **(Division of polynomials)**

52. $\dfrac{72xyz - 48xy - 64xz}{-8x}$

$\dfrac{72xyz}{-8x} - \dfrac{48xy}{-8x} - \dfrac{64xz}{-8x}$

$-9yz + 6y + 8z$ **(Division of polynomials)**

53. $\dfrac{-16xy + 12xz - 24xw}{4x}$

$\dfrac{-16xy}{4x} + \dfrac{12xz}{4x} - \dfrac{24xw}{4x}$

$-4y + 3z - 6w$ **(Division of polynomials)**

54. $\dfrac{32x^2y + 48x^3y^2 - 56xy^3}{-6y}$

$\dfrac{32x^2y}{-6y} + \dfrac{48x^3y^2}{-6y} - \dfrac{56xy^3}{-6y}$

$-\dfrac{16x^2}{3} - 8x^3y + \dfrac{28xy^2}{3}$ **(Division of polynomials)**

55. $\dfrac{75x^2yz - 15xy^2z - 90xyz^2}{15xyz}$

$\dfrac{75x^2yz}{15xyz} - \dfrac{15xy^2z}{15xyz} - \dfrac{90xyz^2}{15xyz}$

$5x - y - 6z$ **(Division of polynomials)**

56. $x[3 - 2(x-1)(x+1)]$

$x[3 - 2(x(x) + x(1) - 1(x) - 1(1))]$ Multiply the two binomials.

$x[3 - 2(x^2 + x - x - 1)]$

$x[3 - 2(x^2 - 1)]$ Collect like terms.

$x[3 - 2x^2 + 2]$ Distribute –2.

$x[-2x^2 + 5]$ Collect like terms.

$-2x^3 + 5x$ **(Order of operations)**

57. $3x(x+2)(x-3)$

$3x[x(x) + x(-3) + 2(x) + 2(-3)]$ Multiply the two binomials.

$3x[x^2 - 3x + 2x - 6]$

$3x[x^2 - x - 6]$ Collect like terms.

$3x(x^2) - 3x(x) - 3x(6)$ Distribute $3x$.

$3x^3 - 3x^2 - 18x$ **(Order of operations)**

58. $3x[(x+2)+(x-3)]$ Collect like terms.

     $3x[2x-1]$

     $3x(2x)-3x(1)$ Distribute $3x$.

     $6x^2-3x$ **(Order of operations)**

59. $(2x^2-5x)-[(3x-2x^2)-(5x^2-2x)]$

     $(2x^2-5x)-[-7x^2+5x]$

     $2x^2-5x+7x^2-5x$

     $9x^2-10x$ **(Order of operations)**

60. $(x^2-2x+1)-[(2x^2-3)+(-3x^2+7)]$

     $(x^2-2x+1)-[-x^2+4]$

     $x^2-2x+1+x^2-4$

     $2x^2-2x-3$ **(Order of operations)**

61. $(4x+1)(3x-2)-(3x+1)(4x-2)$

     $(12x^2-8x+3x-2)-(12x^2-6x+4x-2)$

     $12x^2-5x-2-12x^2+2x+2$

     $-3x$ **(Order of operations)**

62. $6x(x^2-5)-6x(x^2+5x+5)$

     $6x^3-30x-6x^3-30x^2-30x$

     $-30x^2-60x$ **(Order of operations)**

63. $x-[3x^2y-(x-1)]-[-y(2x^2+y)]$

     $x-3x^2y+x-1+2x^2y+y^2$

     $-x^2y+2x+y^2-1$ **(Order of operations)**

64. $(x^2+4)-[(x^2-5)-(3x^2+1)]$

     $x^2+4-[-2x^2-6]$

     x^2+4+2x^2+6

     $3x^2+10$ **(Order of operations)**

65. $\dfrac{9x^2y + 3x}{3x}$

$\dfrac{9x^2y}{3x} + \dfrac{3x}{3x}$

$3xy + 1$ **(Order of operations)**

66. $5x\,(3x - 2) - (2x - 3)^2$

$15x^2 - 10x - (2x - 3)\,(2x - 3)$

$15x^2 - 10x - [4x^2 - 6x - 6x + 9]$

$15x^2 - 10x - [4x^2 - 12x + 9]$

$15x^2 - 10x - 4x^2 + 12x - 9$

$11x^2 + 2x - 9$ **(Order of operations)**

Grade Yourself

Circle the question numbers that you had incorrect. Then indicate the number of questions you missed. If you answered more than three questions incorrectly, you need to focus on that topic. (If a topic has less than three questions and you had at least one wrong, we suggest you study that topic also. Read your textbook or a review book, or ask your teacher for help.)

Subject: *Polynomials*

Topic	Question Numbers	Number Incorrect
Addition of polynomials	1, 2, 3, 4, 5, 6, 7, 8, 9, 10, 11, 12, 13, 14, 15	
Subtraction of polynomials	16, 17, 18, 19, 20, 21, 22, 23, 24, 25, 26, 27, 28, 29, 30	
Multiplication of polynomials	31, 32, 33, 34, 35, 36, 37, 38, 39, 40, 41, 42, 43, 44, 45	
Division of polynomials	46, 47, 48, 49, 50, 51, 52, 53, 54, 55	
Order of operations	56, 57, 58, 59, 60, 61, 62, 63, 64, 65, 66	